[Cooling Water]

GEOTHERMAL POWER PLANT, HAWAI'I

Geothermal energy is a renewable energy source. Geothermal energy is energy that people get from the earth. People can use this energy to heat and cool their homes or create electricity. At this power plant, heat extracted from the earth is used to heat large amounts of water. This heated water is used to spin special turbine blades that make electricity. This electricity can then be used wherever it is needed. Finally, the water is cooled and recycled through the system.

A geothermal heat pump can extract enough heat from shallow ground anywhere in the world to keep a home warm. However, places with greater energy needs have to collect the higher temperature geothermal energy trapped deep in the earth.

NATIONAL GEOGRAPHIC
SCIENCE

PHYSICAL SCIENCE

NATIONAL
GEOGRAPHIC

School Publishing

PROGRAM AUTHORS

Malcolm B. Butler, Ph.D.

Judith S. Lederman, Ph.D.

Randy Bell, Ph.D.

Kathy Cabe Trundle, Ph.D.

David W. Moore, Ph.D.

Program Authors

MALCOLM B. BUTLER, PH.D.

Associate Professor of Science Education,
University of South Florida, St. Petersburg,
Florida
SCIENCE

JUDITH SWEENEY LEDERMAN, PH.D.

Director of Teacher Education,
Associate Professor of Science Education,
Department of Mathematics and Science Education,
Illinois Institute of Technology, Chicago, Illinois
SCIENCE

RANDY BELL, PH.D.

Associate Professor of Science Education,
University of Virginia, Charlottesville, Virginia
SCIENCE

KATHY CABE TRUNDLE, PH.D.

Associate Professor of Early Childhood Science
Education, The School of Teaching and Learning,
The Ohio State University, Columbus, Ohio
SCIENCE

DAVID W. MOORE, PH.D.

Professor of Education,
College of Teacher Education and Leadership,
Arizona State University, Tempe, Arizona
LITERACY

Program Reviewers

Amani Abuhabsah
Teacher
Dawes Elementary
Chicago, IL

Maria Aida Alanis, Ph.D.
Elementary Science
Instructional Coordinator
Austin Independent
School District
Austin, TX

Jamillah Bakr
Science Mentor Teacher
Cambridge Public Schools
Cambridge, MA

Gwendolyn Battle-Lavert
Assistant Professor of Education
Indiana Wesleyan University
Marion, IN

Carmen Beadles
Retired Science Instructional
Coach
Dallas Independent School
District
Dallas, TX

Andrea Blake-Garrett, Ed.D.
Science Educational Consultant
Newark, NJ

Lori Bowen
Science Specialist
Fayette County Schools
Lexington, KY

Pamela Breitberg
Lead Science Teacher
Zapata Academy
Chicago, IL

Carol Brueggeman
K–5 Science/Math Resource
Teacher
District 11
Colorado Springs, CO

Program Reviewers continued
on page iv.

Acknowledgments

Grateful acknowledgment is given to the authors, artists, photographers, museums, publishers, and agents for permission to reprint copyrighted material. Every effort has been made to secure the appropriate permission. If any omissions have been made or if corrections are required, please contact the Publisher.

Illustrator Credits
All illustrations by Precision Graphics.
All maps by Mapping Specialists.

Photographic Credits
Front Cover Phil Degginger/
www.color-pic.com.

Credits continue on page EM10.

Neither the Publisher nor the authors shall be liable for any damage that may be caused or sustained or result from conducting any of the activities in this publication without specifically following instructions, undertaking the activities without proper supervision, or failing to comply with the cautions contained herein.

The National Geographic Society
John M. Fahey, Jr.,
President & Chief Executive Officer

Gilbert M. Grosvenor,
Chairman of the Board

Copyright © 2011 The Hampton-Brown Company, Inc., a wholly owned subsidiary of the National Geographic Society, publishing under the imprints National Geographic School Publishing and Hampton-Brown.

National Geographic School Publishing
Hampton-Brown
www.myNGconnect.com

Printed in the USA.
RR Donnelley
Jefferson City, MO

ISBN: 978-0-7362-7808-9

11 12 13 14 15 16 17 18 19 20

2 3 4 5 6 7 8 9 10

Miranda Carpenter
Teacher, MS Academy Leader
Imagine School
Bradenton, FL

Samuel Carpenter
Teacher
Coonley Elementary
Chicago, IL

Diane E. Comstock
Science Resource Teacher
Cheyenne Mountain School
District
Colorado Springs, CO

Kelly Culbert
K–5 Science Lab Teacher
Princeton Elementary
Orange County, FL

Karri Dawes
K–5 Science Instructional
Support Teacher
Garland Independent
School District
Garland, TX

Richard Day
Science Curriculum Specialist
Union Public Schools
Tulsa, OK

Michele DeMuro
Teacher/Educational
Consultant
Monroe, NY

Richard Ellenburg
Science Lab Teacher
Camelot Elementary
Orlando, FL

Beth Faulkner
Brevard Public Schools
Elementary Training Cadre,
Science Point of Contact,
Teacher, NBCT
Apollo Elementary
Titusville, FL

Kim Feltre
Science Supervisor
Hillsborough School District
Newark, NJ

Judy Fisher
Elementary Curriculum
Coordinator
Virginia Beach Schools
Virginia Beach, VA

Anne Z. Fleming
Teacher
Coonley Elementary
Chicago, IL

Becky Gill, Ed.D.
Principal/Elementary Science
Coordinator
Hough Street Elementary
Barrington, IL

Rebecca Gorinac
Elementary Curriculum Director
Port Huron Area Schools
Port Huron, MI

Anne Grall Reichel Ed. D.
Educational Leadership/
Curriculum and Instruction
Consultant
Barrington, IL

Mary Haskins, Ph.D.
Professor of Biology
Rockhurst University
Kansas City, MO

Arlene Hayman
Teacher
Paradise Public School District
Las Vegas, NV

DeLene Hoffner
Science Specialist, Science
Methods Professor,
Regis University
Academy 20 School District
Colorado Springs, CO

Cindy Holman
District Science Resource
Teacher
Jefferson County Public Schools
Louisville, KY

Sarah E. Jesse
Instructional Specialist for
Hands-on Science
Rutherford County Schools
Murfreesboro, TN

Dianne Johnson
Science Curriculum Specialist
Buffalo City School District
Buffalo, NY

Kathleen Jordan
Teacher
Wolf Lake Elementary
Orlando, FL

Renee Kumiega
Teacher
Frontier Central School District
Hamburg, NY

Edel Maeder
K–12 Science Curriculum
Coordinator
Greece Central School District
North Greece, NY

Trish Meegan
Lead Teacher
Coonley Elementary
Chicago, IL

Donna Melpolder
Science Resource Teacher
Chatham County Schools
Chatham, NC

Melissa Mishovsky
Science Lab Teacher
Palmetto Elementary
Orlando, FL

Nancy Moore
Educational Consultant
Port Stanley, Ontario, Canada

Melissa Ray
Teacher
Tyler Run Elementary
Powell, OH

Shelley Reinacher
Science Coach
Auburndale Central
Elementary
Auburndale, FL

Kevin J. Richard
Science Education Consultant,
Office of School Improvement
Michigan Department of
Education
Lansing, MI

Cathe Ritz
Teacher
Louis Agassiz Elementary
Cleveland, OH

Rose Sedely
Science Teacher
Eustis Heights Elementary
Eustis, FL

Robert Sotak, Ed.D.
Science Program Director,
Curriculum and Instruction
Everett Public Schools
Everett, WA

Karen Steele
Teacher
Salt Lake City School District
Salt Lake City, UT

Deborah S. Teuscher
Science Coach and
Planetarium Director
Metropolitan School District
of Pike Township
Indianapolis, IN

Michelle Thrift
Science Instructor
Durrance Elementary
Orlando, FL

Cathy Trent
Teacher
Ft. Myers Beach Elementary
Ft. Myers Beach, FL

Jennifer Turner
Teacher
PS 146
New York, NY

Flavia Valente
Teacher
Oak Hammock Elementary
Port St. Lucie, FL

Deborah Vannatter
District Coach, Science
Specialist
Evansville Vanderburgh School
Corporation
Evansville, IN

Katherine White
Science Coordinator
Milton Hershey School
Hershey, PA

Sandy Yellenberg
Science Coordinator
Santa Clara County Office
of Education
Santa Clara, CA

Hillary Zeune de Soto
Science Strategist
Lunt Elementary
Las Vegas, NV

PHYSICAL SCIENCE

CONTENTS

TECHTREK
myNGconnect.com

Student eEdition

Vocabulary Games

Digital Library

Enrichment Activities

CHAPTER

3

TECHTREK
myNGconnect.com

Student eEdition   Vocabulary Games   Digital Library   Enrichment Activities

PHYSICAL SCIENCE

well as how those objects interact with each other. Physical science includes the study of matter, motion and forces, and many kinds of energy, including light and electricity. People who study how all of these things work together are called physical scientists.

You will learn about these aspects of physical science in this unit:

HOW CAN YOU DESCRIBE MATTER, MIXTURES, AND SOLUTIONS?

Matter is anything that has mass and takes up space. Physical scientists study all of the different properties of matter. These include size, shape, color, texture, hardness, and temperature, as well as mass and volume. Physical scientists study the atom and the makeup of matter at the smallest level as well.

HOW CAN MATTER CHANGE?

Physical scientists study how matter changes. Matter can undergo a physical change, such as a cut or a tear. Matter can also undergo a chemical change, such as burning or rusting. Often, new matter results from these changes.

HOW DO YOU DESCRIBE FORCE AND THE LAWS OF MOTION?

Physical scientists study how objects move. They also study the forces that act on objects, and how those objects respond to different kinds of forces. Isaac Newton was a famous physical scientist who described his observations of the effects of force on objects in what is now known as the laws of motion.

WHAT ARE SIMPLE MACHINES?

Simple machines make our lives easier in many ways. Simple machines change the direction of the force or spread the force needed over a greater distance. Physical scientists study simple machines to make the tools we use safer and more efficient.

HOW DO YOU DESCRIBE DIFFERENT FORMS OF ENERGY?

Physical scientists study energy in all of its forms. They also learn about how energy can cause changes in the physical world. The different kinds of energy include light, sound, electrical, and mechanical.

HOW DOES ELECTRICAL ENERGY FLOW AND TRANSFORM?

Many objects in our world depend on electrical energy for the energy needed to function. Physical scientists study which objects electricity flows through easily, and which objects it does not. They also study how electricity can change into other forms of energy, such as light and sound.

MEET A SCIENTIST

Thomas Taha Rassam Culhane:
Urban Planner

T.H., as he likes to be called, is an urban planner and National Geographic Emerging Explorer. T.H. and his team work in some of the poorest neighborhoods of Cairo, Egypt, and elsewhere to build and install rooftop solar water heaters, kitchen-waste-to-biogas digesters and other devices to meet people's everyday needs for cooking fuel, electricity, and clean water.

T.H. explains. "The technology we create and install with the people is completely carbon dioxide free, it contributes nothing to global warming. If people don't have access to enough water, it becomes a serious health issue. And when women spend all their time tending stoves to heat water, how can they go to school or get ahead?" T.H.'s grasp of these daily challenges is personal; he and his wife once moved to a slum apartment themselves to gain firsthand experience.

T.H. uses kitchen scraps and waste water that would have normally been put in the trash or down the drain to create cooking fuel in a biogas digester. The digester provides enough fuel for a family to cook their meals.

 4

6
7

After reading Chapter 1, you will be able to:

- Classify matter as a solid, a liquid, or a gas, as it exists at room temperature, using physical properties. **PROPERTIES OF MATTER**

- Recognize that particles are always in motion, with the smallest motion in solids and the largest motion in gases. **PROPERTIES OF MATTER**

- Recognize that a substance has characteristic properties such as density, boiling point, freezing point, and solubility, all of which are independent of the mass or volume of the sample. **PROPERTIES OF MATTER**

- Recognize that mass is a measure of the amount of matter in an object. **MASS AND VOLUME**

- Show that equal volumes of different substances usually have different masses. **MASS AND VOLUME**

- Understand that all matter is made up of atoms, which are far too small to see. **ATOMS AND ELEMENTS**

- Understand that mixtures are physical combinations of materials and can be separated by physical means. **MIXTURES**

- Describe the properties of mixtures and solutions, including concentration and saturation. **MIXTURES, SOLUTIONS**

- Recognize that the rate of solution can be affected by the size of the particles, stirring, temperature, and the amount of solute already dissolved. **SOLUTIONS**

- Science in a Snap! Understand that all matter is made up of atoms, which are far too small to see directly through a microscope. **ATOMS AND ELEMENTS**

HOW CAN YOU DESCRIBE MATTER, MIX

Look around you. What do you see? You see matter. Matter is everywhere. You can see it in the forms of liquids, solids, and gases. What matter can you observe in this picture?

Solids and liquids are easy to see. The diver's scuba gear is a solid. The ocean water is a liquid. But what about gases? The bubbles you can see around the diver are gas.

TECHTREK
myNGconnect.com

Student
eEdition

Vocabulary
Games

Digital
Library

Enrichment
Activities

TURES, AND SOLUTIONS?

SCIENCE VOCABULARY

mass (MAS)

Mass is the amount of matter in an object. (p. 14)

The items on the pan balance have equal mass.

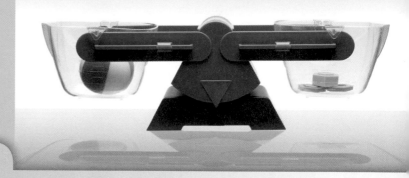

volume (VOL-yum)

Volume is the amount of space something takes up. (p. 16)

A beaker is a tool that measures the volume of a liquid.

atom (A-tum)

An **atom** is the smallest piece of matter that can still be identified as that matter. (p. 18)

This is a model of an atom.

my
Science Vocabulary

atom
(A-tum)

mass
(MAS)

mixture
(MIKS-chur)

solution
(so-LŪ-shun)

volume
(VOL-yum)

TECHTREK
myNGconnect.com

Vocabulary
Games

mixture (MIKS-chur)

A **mixture** is two or more kinds of matter put together. (p. 26)

This mixture of nuts and bolts can easily be sorted by color and size.

solution (so-LŪ-shun)

A **solution** is a mixture of two or more kinds of matter evenly spread out. (p. 30)

The water in a swimming pool is a solution of water and chlorine.

Properties of Matter

How do scientists define *matter*? Matter is anything that has mass and takes up space. Scientists describe and compare matter based on properties, or qualities. State, color, texture, temperature, mass, and volume are some of matter's properties.

State Solids, liquids, and gases are called states of matter. While ice is a solid, for example, water is a liquid.

Some matter, such as a block of wood, has a definite shape and volume. A piece of wood is a solid. The tightly packed particles in all solids give them their shapes.

If the particles in the matter are able to move around, the matter may be liquid. Liquids such as water, juice, and milk have definite volumes and take the shapes of their containers.

If you've ever seen a balloon floating in the air, you've seen an object filled with gas. Gases have no definite shape or volume. Instead, their quickly moving particles spread out to fill a space.

Particles in solids are packed tightly together.

Particles in liquids are able to move, though they bump into each other.

Particles in gases are spread far apart, move rapidly, and sometimes bump into each other.

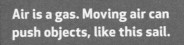
Air is a gas. Moving air can push objects, like this sail.

The ocean is a liquid.

The surfboard is a solid.

11

Color and Texture Color and texture are two properties of matter that scientists describe and compare. Solids, liquids, and gases may all have color. Solids and liquids can have texture.

One property of matter is color. *Red, blue,* and *yellow* are just a few words that describe color.

If you've ever run your hand over sandpaper or fur, you've felt texture. Textures, such as rough, smooth, bumpy, and furry, are properties you can feel.

Flowers come in a variety of colors.

The texture of a tennis ball can be described as soft, furry, or fuzzy. A baseball's texture is smooth and leathery. Basketballs have a bumpy or rough texture.

Temperature

Your tea is too hot, so you add ice to make it cooler. You are changing the temperature of your drink. Temperature is a measure of how hot or cold an object is. How can scientists measure and describe the property of temperature? They use thermometers and two different scales. The Fahrenheit scale is the scale used most often in the United States. People in other countries, and scientists around the world, measure temperature using the Celsius scale.

Water freezes to form ice at 0°C (32°F).

Iron ore is heated to 871°C (1,600°F) to create steel.

Before You Move On

1. What properties do scientists use to describe and compare matter?
2. How would you describe the color, texture, and temperature of your science book to a friend?
3. **Analyze** How could properties such as color, texture, and temperature help scientists distinguish between two different samples of matter?

Mass and Volume

Mass Some big objects, such as trucks, have a lot of matter. Smaller objects, such as toy trucks, have much less matter. How can you measure the amount of matter in an object? You can figure out its mass. Mass is the amount of matter in an object. Mass is measured using metric units: grams and kilograms. To find the mass of an object, use a pan balance. Place the object on one side of the balance. Add gram weights to the other side until the sides are balanced. Add the sum of the gram weights to find the mass of the object.

Look at the picture below. What do you notice about the two sides of the pan balance? What does this tell you about the mass of the ball?

TECHTREK
myNGconnect.com

You can use a balance like this to compare the masses of objects.

Digital Library

14

Mass is different from weight. Your weight is a measure of gravity's pull on you. On the moon you would weigh less than on Earth because the moon's gravitational pull on your body is less. Your weight on the moon would be about one-sixth of your weight on Earth. On Jupiter, which has a larger gravitational pull, you would weigh more, about 2.5 times more than you weigh on Earth. But you would still have the same mass in all three places. The only thing that has changed is gravity's pull.

The smaller force of gravity on the moon would affect your weight. However, your mass would remain the same.

The force of gravity on Earth affects your weight but not your mass.

Volume Matter takes up space. The amount of space taken up by an object is its **volume** . You can measure the volume of solids, liquids, and gases.

Calculating Volume If you want to find the volume of an object such as a block, you can measure it and then use math. Measure the length, width, and height and then multiply the three numbers.

How would you measure the volume of a marble or rock? Fill a beaker of water and note the volume. Then, place the object in the beaker. Note the new reading. To find the volume of the solid object, find the difference between the two volumes.

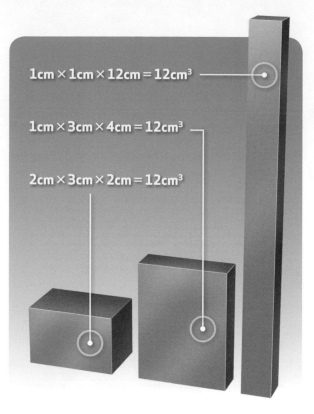

$1cm \times 1cm \times 12cm = 12cm^3$

$1cm \times 3cm \times 4cm = 12cm^3$

$2cm \times 3cm \times 2cm = 12cm^3$

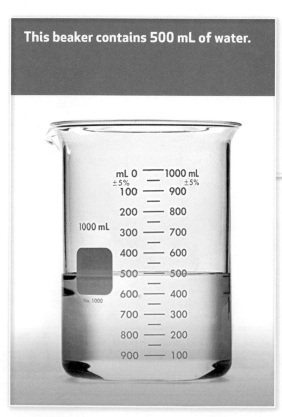

This beaker contains 500 mL of water.

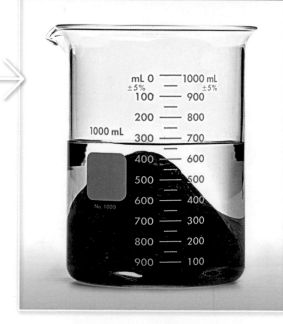

The water in the beaker now measures 675 mL. The volume of the rock is 175 mL.

Liquids have a definite volume, but they take the shape of their containers. Look at the picture. The volume of the liquid is the same in both containers, but the containers are different shapes.

Volume of Gases Gases spread out to fill their containers. The volume of gas is the same as the size of its container. The volume of air in a liter bottle, for example, would be one liter.

Mass and Volume If you had one liter of air and one liter of water, their masses would be different. The same volumes of different substances may have different masses.

Enrichment Activities

Because liquid takes the shape of its container, the volume of liquid looks different in each container.

One propane container has the same volume as nine two-liter bottles.

Before You Move On

1. What methods can you use to measure the volume of a solid?
2. How are the volumes of solids, liquids, and gases different?
3. **Infer** What would happen to your weight if the force of gravity on Earth suddenly grew stronger? How about your mass?

Atoms and Elements

Think about your favorite book. What is the smallest part that makes up the book? Writers use small pieces—letters—to create words, sentences, paragraphs, and, eventually, the story. Like individual letters, atoms are the smallest parts that make up matter. Everything in our world— you, this book, Earth itself—is made up of tiny atoms.

There are over 100 different kinds of atoms in the universe. Every atom has three parts: protons, electrons, and neutrons. Atoms are too small to be seen.

This balloon will soon be full of atoms of gas!

The protons in an atom are in the nucleus. A neutral atom has the same number of electrons as protons.

nucleus ———— neutron
proton
electron

Protons exist in the center, or nucleus, of the atom. Protons have a positive electrical charge. The number of protons in the nucleus identifies the atom. For example, helium atoms have two protons.

Neutrons exist with protons in the nucleus of the atom. Neutrons have no electrical charge.

Electrons move around the nucleus of an atom. Electrons have a negative electrical charge. In a neutral atom, the number of electrons in an atom is equal to the number of protons in the nucleus.

Science in a Snap! Close-Up

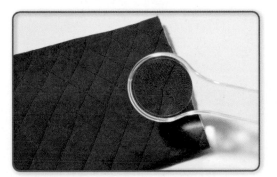

Use a hand lens to closely examine fabric, such as on your shirt or sweater.

Describe what you see.

How did your view of the fabric change with the hand lens?

Elements Elements are the building blocks of all matter. Elements cannot be broken into different parts. Every atom is an atom of a particular element. There are over 100 known elements in the universe. About 92 of these elements occur in nature. Scientists make the rest. Four elements—carbon (C), oxygen (O), hydrogen (H), and nitrogen (N)—make up about 96 percent of living matter.

The elements have been organized into a table called the Periodic Table of the Elements. Each element appears on the table in order based on its atomic number, the number of protons in its nucleus. Each element has an abbreviation to make it easy to identify. Elements that are alike are placed together in the table.

IRON IN OUR UNIVERSE

Iron is abundant in the universe. It is found in the sun, in stars, and in Earth.

Some plants such as spinach are rich in iron.

Iron is a part of blood cells that helps deliver oxygen throughout the body.

THE **PERIODIC TABLE** OF THE ELEMENTS

1 H Hydrogen 1.00794																	2 He Helium 4.003
3 Li Lithium 6.941	4 Be Beryllium 9.012182											5 B Boron 10.811	6 C Carbon 12.0107	7 N Nitrogen 14.00674	8 O Oxygen 15.9994	9 F Fluorine 18.998403	10 Ne Neon 20.1797
11 Na Sodium 22.989770	12 Mg Magnesium 24.3050											13 Al Aluminum 26.981538	14 Si Silicon 28.0855	15 P Phosphorus 30.978706	16 S Sulfur 32.066	17 Cl Chlorine 35.492	18 Ar Argon 39.948
19 K Potassium 39.0983	20 Ca Calcium 40.078	21 Sc Scandium 44.955910	22 Ti Titanium 47.867	23 V Vanadium 50.9415	24 Cr Chromium 51.9961	25 Mn Manganese 54.938049	26 Fe Iron 55.845	27 Co Cobalt 58.933200	28 Ni Nickel 58.6934	29 Cu Copper 63.546	30 Zn Zinc 65.39	31 Ga Gallium 69.723	32 Ge Germanium 72.61	33 As Arsenic 74.92160	34 Se Selenium 78.96	35 Br Bromine 79.904	36 Kr Krypton 83.80
37 Rb Rubidium 85.4678	38 Sr Strontium 87.62	39 Y Yttrium 88.90585	40 Zr Zirconium 91.224	41 Nb Niobium 92.90638	42 Mo Molybdenum 95.94	43 Tc Technetium (98)	44 Ru Ruthenium 101.07	45 Rh Rhodium 102.90330	46 Pd Palladium 106.42	47 Ag Silver 107.8682	48 Cd Cadmium 112.411	49 In Indium 114.818	50 Sn Tin 118.710	51 Sb Antimony 121.760	52 Te Tellurium 127.60	53 I Iodine 126.90447	54 Xe Xenon 131.29
55 Cs Cesium 132.90545	56 Ba Barium 137.327	57 La Lutetium 138.9055	72 Hf Halfnium 178.49	73 Ta Tantalum 180.9479	74 W Tungsten 183.84	75 Re Rhenium 186.207	76 Os Osmium 190.23	77 Ir Iridium 192.217	78 Pt Platinum 195.078	79 Au Gold 196.96655	80 Hg Mercury 200.59	81 Tl Thallium 204.3833	82 Pb Lead 207.2	83 Bi Bismuth 208.98038	84 Po Polonium (209)	85 At Astatine (210)	86 Rn Radon (222)
87 Fr Francium (223)	88 Ra Radium (226)	89 Ac Lawrencium (227)	104 Rf Rutherfordium (261)	105 Db Dubnium (262)	106 Sg Seaborgium (263)	107 Bh Bohrium (262)	108 Hs Hassium (265)	109 Mt Meitnerium (266)	110 (269)	111 (272)	112 (277)	113	114				

58 Ce Cerium 140.116	59 Pr Praseodymium 140.90765	60 Nd Neodymium 144.24	61 Pm Promethium (145)	62 Sm Samarium 150.36	63 Eu Europium 151.964	64 Gd Gadolinium 157.25	65 Tb Terbium 158.92534	66 Dy Dysprosium 162.50	67 Ho Holmium 164.93032	68 Er Erbium 167.26	69 Tm Thulium 168.93421	70 Yb Ytterbium 173.04	71 Lu Lanthanum 174.967
90 Th Thorium 232.0381	91 Pa Protactinium 231.03588	92 U Uranium 238.0289	93 Np Neptunium (237)	94 Pu Plutonium (244)	95 Am Americium (243)	96 Cm Curium (247)	97 Bk Berkelium (247)	98 Cf Californium (251)	99 Es Einsteinium (252)	100 Fm Fermium (257)	101 Md Mendelevium (258)	102 No Nobelium (259)	103 Lr Lawrencium (262)

FLUORINE: used in toothpaste

BERYLLIUM: found in emeralds

IRON: used in steel construction

Molecules Letters are small parts that make up bigger parts, words. In the same way, atoms are small parts that make up bigger parts called molecules. Atoms can combine in different ways to make different kinds of molecules.

Every living and nonliving thing in the universe is made up of molecules. A molecule can be made of two atoms or of hundreds of millions of atoms.

Scientists use special symbols that represent atoms and molecules. You may have seen the symbol for water: H_2O. The symbol means that every molecule of water is made of two hydrogen atoms (H) and one oxygen atom (O).

Molecules are formed when atoms bond with each other. Atoms bond with other atoms by giving, taking, or sharing their electrons.

One drop of water has 1,670,000,000,000,000,000,000 water molecules. That's a huge number!

COMMON **MOLECULES**

A water molecule is made up of one atom of oxygen and two atoms of hydrogen.

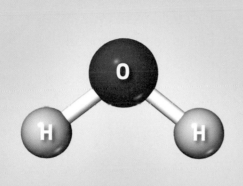

Human beings produce carbon dioxide every time they breathe. Plants use carbon dioxide to live and grow.

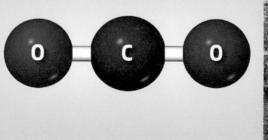

Compounds A compound is a molecule that contains at least two different elements. The combination of an oxygen atom and two hydrogen atoms creates a compound—water (H_2O). Because a water molecule contains two different elements—hydrogen and oxygen—it is a compound. We call each small "piece" of water a water molecule.

When elements combine to form a compound, the new compound has properties that are different from those of the elements that make it up. For example, sodium (Na) is a metal. Chlorine (Cl) is a gas at room temperature. Both sodium and chlorine are poisonous on their own. When the two are combined in a chemical reaction, they form the compound sodium chloride (NaCl), or salt. Salt is neither a metal nor a gas. It is safe enough to eat!

Sodium chloride is the chemical name for salt.

William Gerwick collects natural compounds in search of new medicines.

Many compounds occur naturally on earth. Carbon dioxide (CO_2) is a compound of carbon and oxygen. You can find carbon dioxide in the air. We breathe it out. We add carbon dioxide to our drinks to make them fizz. When we bake cakes and breads, carbon dioxide makes them rise.

TECHTREK
myNGconnect.com

Digital Library

SOME COMMON **COMPOUNDS**

COMMON NAME	CHEMICAL NAME	
CHALK	Calcium Sulfide	
VINEGAR	Acetic Acid	
RUST	Iron Oxide	

Before You Move On

1. What are the parts of an atom?
2. How are molecules and compounds alike and different?
3. **Infer** How are chlorine and sodium different from the salt they form together?

Mixtures

Look at the nuts and bolts on this page. They are not all alike. They are a **mixture**, two or more things that are combined. A mixture can be separated by its properties, such as size, color, and shape. What properties would you use to separate the mixture of nuts and bolts?

Mixtures are made up of different amounts of two or more substances. There do not have to be equal amounts, and the amount of each different substance can change. Each part of a mixture keeps its original properties. The parts of a mixture can be easily separated.

TECHTREK
myNGconnect.com

You can separate these nuts and bolts by their properties.

Digital Library

You can separate the mixture of nuts and bolts by color and size. There are many ways to separate mixtures according to their properties.

Look at the mixture on this page. Some of the objects are magnetic, and others are not. You could use a magnet to separate some parts of the mixture. The paper clips, pins, and screws separate from the rubber bands and plastic figures. Once the magnetic items are sorted from the ones that do not attract magnets, how could you separate items in each group based on their properties?

The magnets attracts all of the magnetic objects. The non-magnetic objects do not contain iron. They are left on the table.

Types of Mixtures

• Look at the picture of the mud. See the liquid and the solid particles sticking to the girl's hands? Mud is a **solid/liquid** mixture. Cake batter and lemonade are other mixtures of liquids and solids.

• A **liquid/liquid mixture** is a combination of two liquid ingredients. Oil and vinegar salad dressing is a mixture that contains only liquids. Gasoline is this type of mixture too.

Oil and vinegar salad dressing is a liquid/liquid mixture. The oil is on top of the vinegar in the photograph.

- Liquid matter and gases can combine to form **liquid/gas** mixtures. Soda water and bubbles found in foaming baths and dishwater are examples of mixtures that contain both liquids and gases.

Bubbles containing gas form in this dishwater.

- The cereal is a mixture of flakes, nuts, and dried fruits. **Solid/solid** mixtures contain all solid parts. Soil is another mixture made only of solids.

Before You Move On

1. What are the properties of a mixture?
2. You are holding a glass of orange juice with ice cubes floating in it. What kind of mixture are you holding?
3. **Analyze** If you found an interesting mixture, what properties would you look for to try to separate it into its individual components?

Solutions

If you have ever swum in a pool, you have swum in a solution! A **solution** is a mixture of two or more types of matter evenly spread out and not easily separated. Every sample of the solution will contain the exact same parts, properties, and appearance. A solution is made up of two parts: a solute and a solvent.

In a swimming pool, liquid chlorine is added to water to make the water safe and clean for swimming. The water is the solvent—there is more water than liquid chlorine. The liquid chlorine is dissolved in the water to make a solution. That makes the chlorine the solute.

Pool water is an example of a solution made up of two liquids: water and liquid chlorine.

Solutions can also contain materials in different states. Juice made from a powder mix, for example, is a solution of a liquid solvent (water) and a solid solute (juice mix). Solubility is matter's ability to dissolve in a liquid. Look at the images of solids below. Which do you think are soluble in water? Which are not?

Factors that affect how fast a solid dissolves include:

• size: smaller solids dissolve faster.
• temperature: solids dissolve faster at higher temperatures.
• stirring: solids dissolve faster when the solution is stirred.

When salt or sugar is added to water, it dissolves and evenly distributes throughout the water, creating a solution. When you look at the solution, you cannot see the particles of salt or sugar. When you add sand or a coin to water, you have created a mixture. The sand and coin do not dissolve and distribute through the water. At some points in the mixture, you will find just water. At other points, you will find sand or a coin.

WHICH OF THESE SOLIDS ARE SOLUBLE IN WATER?

salt

sand

quarter

sugar

Concentration Look at the three glasses of juice on the next page. They are all made with water and juice mix. What do you notice about them? They are different colors, because each of them has a different amount of juice mix. The juice glasses have solutions with different concentrations. The concentration of a solution is the amount of solute that is dissolved in the solution.

The juice on the left has a low concentration. It has a small amount of juice mix compared to the amount of water in the glass. The second picture shows a medium concentration, while the third picture shows a high concentration of juice mix.

Scientists can classify solutions based on their concentrations. A concentrated solution has a lot of solute in it. The juice on the far right is the most concentrated solution.

The water in this Florida swamp contains a high concentration of dirt and other materials.

Dilute means "thinned." The middle glass of juice could be described as a dilute solution. The glass of juice on the left, the weakest solution, would be described as very dilute.

Sometimes solute is added to a solvent until the solvent cannot dissolve any more solute. If you kept adding juice mix to the water, for example, eventually the water would stop dissolving the mix. Extra juice mix would stay solid, sinking to the bottom or floating to the top of the liquid.

Before You Move On

1. What are the parts of a solution?
2. How does the amount of a solute in a solvent affect the solution's concentration?
3. **Infer** What do you think happens when Earth's atmosphere becomes saturated with water vapor?

A NATURAL SOLUTION

The Dead Sea is one of the smallest seas in the world, less than 80.5 km (50 miles) long, ranging from 3.2 km (2 m) to 17.7 km (11 m) wide. Located between Jordan and Israel in southwestern Asia, the Dead Sea is surrounded by land. The name "Dead Sea" is a gentle translation from its Hebrew name, "Yam ha Maved," which literally means "Killer Sea." At 418 meters (1372 feet) below sea level, the Dead Sea has the lowest water surface in the world.

The Dead Sea is the saltiest body of water in the world. The Jordan River and some smaller streams feed the surface of the Dead Sea. The fresh water from these sources is quickly evaporated in the hot desert climate. At the surface, the water is the least salty. As the water gets deeper, it gets much saltier. At about 40 meters (130 feet) deep, the Dead Sea is about ten times as salty as the oceans. Deeper, below 90 m (300 ft) the sea is completely saturated. The water at this point cannot hold any more salt. The salt settles in drifts at the bottom of the sea.

The Dead Sea is in Israel. The concentration of salt is so high that people float on the water.

Salt collects along the shore of the Dead Sea.

No rivers or streams lead out of the Dead Sea to the sea. With no way for the water to escape, it just becomes saltier and saltier. How? The fresh water that feeds into it settles on the surface and quickly evaporates, leaving behind minerals. These minerals sink into the water below, increasing the concentration of salt in the water.

No animals live in or around the Dead Sea. The environment is too harsh. A few bacteria and algae have adapted to its severe conditions. Fish that accidentally swim into the sea from the Jordan River or other streams die instantly. Their bodies become covered in salt crystals and they wash up on shore. The fish become part of the white, salt-covered landscape that surrounds the sea.

The concentration of salt in the Dead Sea is so high that the water is much denser than ocean or fresh water. This allows humans to simply float at the surface like corks. Because of the dense water, it is actually difficult to swim in the sea. Most people who visit simply lie back in the water, float around, and read. People with arthritis and skin problems also visit the Dead Sea. The seawater helps them feel better.

Conclusion

Matter is anything that has mass, takes up space, and occurs as a solid, liquid, or gas. Scientists look at the properties of matter when describing or comparing. These properties include color, texture, temperature, mass, and volume. Matter is made up of particles called atoms. Matter can combine to form mixtures and solutions.

Big Idea Everything in the universe is made up of matter, which is made up of atoms.

The helium inside this balloon is matter in the gas state.

Water is matter in the liquid state.

Bricks are matter in the solid state.

Vocabulary Review

Match the following terms with the correct definition.

A. mass

B. volume

C. atom

D. mixture

E. solution

1. The amount of space something takes up
2. The smallest piece of matter that can still be identified as that matter
3. Two or more kinds of matter put together
4. The amount of matter in an object
5. A mixture of two or more kinds of matter evenly spread out

Big Idea Review

1. **List** List at least three properties you can use to describe matter.

2. **Describe** How is weight different from mass?

3. **Compare and Contrast** How is a solution of salt and water different from a mixture of salt and gravel?

4. **Cause and Effect** You and a friend are making a pitcher of lemonade. The directions call for 1 liter of water, but you add 2 liters by mistake. How will that affect the properties of the solution?

5. **Analyze** If you had a mixture of marbles, how could you separate them into groups based on their properties?

6. **Generalize** You have an unknown solid substance. It is shaped so that it cannot be measured with a ruler. What properties can you use to describe the object? How can you measure it?

Write About the Properties of Matter, Mixtures, and Solutions

Explain Is this a mixture or a solution? How do you know? Explain how you could separate these items.

PHYSICAL SCIENCE EXPERT: CHEMIST

Have you ever thought about the way a plant makes food? It takes energy from the sun to make the food. The food is a new kind of matter that it needs to live and grow.

Tehshik Yoon is using this same idea to make new kinds of matter. He is a teacher at a university. Dr. Yoon and his research group ask a lot of questions. They want to know the best way to capture sunlight. They are also trying to find out how to use sunlight to make new kinds of matter.

Dr. Yoon teaches students about matter.

TECHTREK
myNGconnect.com

Digital Library

Dr. Yoon uses a special machine to help him make new kinds of matter.

Dr. Yoon has always been interested in building things. As a child, he spent his time building structures out of blocks. Now he builds structures that are so small you cannot see them!

Dr. Yoon's work is very important in many fields, but it is especially important in making medicines. Now people will be able to make medicines using sunlight, which is a clean source of energy. Many chemical plants have to burn oil and use electricity in order to make medicines. This can be harmful to the environment. Using sunlight to make the same medicines costs less money and causes less pollution. In the future, this research may also lead to finding new medicines.

BECOME AN EXPERT

Matter: The Uses of Matter

Matter is everywhere. Everything you see, feel, taste, and smell is made up of matter. Many forms of matter are found naturally in the earth. Matter that is found in the earth is used to make many different things. Cars, video games, and even artificial body parts all began as matter from the earth.

Matter is found in the earth in all three states: solid, liquid, and gas. These materials are removed from the earth through mining and drilling. It takes well-trained people to decide whether or not to dig for matter. It is a difficult and expensive process.

TECHTREK
myNGconnect.com

Student
eEdition

Digital
Library

These machines remove oil from the earth.

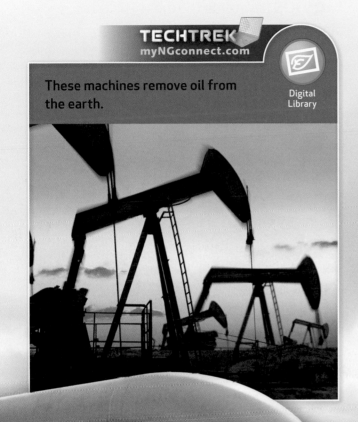

Energy such as gasoline, heating gas, and electricity come from natural resources in the earth. Natural resources are things such as trees, water, coal, oil, natural gas, and soil. We find natural resources in the environment and use them to create useful products.

Natural gas flows through these huge pipes.

41

Solids Many solids occur naturally in the earth. They are found by themselves or as mixtures of different materials. Iron, silver, and gold are all metals that are found in the earth. Sometimes the gold you find in the earth is not pure gold. It's part of a **mixture** called ore.

Liquids Water is the most plentiful liquid on Earth. It covers 70 percent of Earth's surface. We use water for drinking, irrigation, and energy. Oil is another liquid pulled from the earth.

Gases Natural gas comes from deep below Earth's surface. Natural gas is one of the most used forms of energy on Earth. Natural gas is made up mostly of methane. Methane is a compound that contains one **atom** of carbon and four atoms of hydrogen. The air in our atmosphere has energy. Wind power can generate electricity.

Gold is solid.

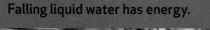
Falling liquid water has energy.

People harvest blowing air, a solution of gases, to use its energy.

mixture
A **mixture** is two or more kinds of matter put together.

atom
An **atom** is the smallest piece of matter that can still be identified as that matter.

Coal Fossil fuels started as the remains of plants and animals. Over millions of years, heat and pressure broke the remains down into much of what we use today for energy. There are three kinds of fossil fuels: coal, oil, and natural gas. Coal is found in the earth as a solid. It is a rock made up mostly of carbon, hydrogen, and oxygen. Of the three types of fossil fuels, coal is the most plentiful and least expensive. People around the world use coal to make electricity. Mining is the process of removing coal from the ground.

Using coal to create energy has some risks. In areas where coal is stripped off the surface of the land, the land is sometimes destroyed. Mining coal and burning coal also create air pollution. The coal industry and national and state governments are working to make the level of pollution go down and to take better care of the land around mines. These practices can make coal mining less risky and dangerous.

U.S.A. COAL-PRODUCING STATES

Wyoming

Pennsylvania
West Virginia
Virginia
Kentucky

The United States has the largest supply of coal in the world. Large coal-producing states include Kentucky, Wyoming, Pennsylvania, Virginia, and West Virginia.

Diamonds Diamonds are a natural material. They are formed deep inside the earth. Diamonds have been discovered in more than 35 countries around the world. The **mass** of a diamond is measured in carats. The largest diamond ever mined was over 3,000 carats. It was discovered in 1905 in South Africa.

Gold Gold is a precious metal that is found in a solid state within the earth. Long ago, people could find gold by sifting through rocks in a river or hitting rocks with a pickaxe. Today, gold is found in much smaller amounts. To mine gold, miners blast large craters in the ground and remove huge amounts of ore. Ore is a mixture of gold and rocks, such as quartz. The ore is crushed into a powder and mixed with water. Then a chemical **solution** is added to separate out the gold.

Almost half the world's gold can be found in South Africa.

mass

Mass is the amount of matter in an object.

solution

A **solution** is a mixture of two or more kinds of matter evenly spread out.

Oil Oil, or petroleum, is a liquid fossil fuel found deep in the earth. Oil is a mixture made mostly of liquid carbon and hydrogen, along with other materials. Geologists are scientists who study Earth. Geologists locate oil underground. After a site has been found and cleared, drilling begins. A well is drilled and prepared so no earth will fall into the hole. Then, a pump is placed in the well. An electric motor moves a lever up and down, creating suction. Like drinking with a straw, suction draws the oil up the pump.

Oil that is straight from the ground and has not been cleaned up or processed is called crude oil. Oil in the ground can have different colors and textures. Coal can be burned as it is when it comes out of the ground. Oil, however, must be refined before it can be burned. *Refined* means that the oil has been cleaned up and the extra matter has been removed.

Oil wells create suction that draws oil from the earth.

Refining Oil Crude oil must be refined before we can use it. When crude oil, or petroleum, is refined, its many parts are separated with heating. The different materials in the oil boil at different temperatures. As each part boils, it changes to a gas. As each gas cools, it turns back into a liquid and is collected. We use each part of the oil for different things. The most common petroleum product is gasoline.

Some people estimate that there are over 6,000 products made from petroleum. The chart below shows just a few of the products made from petroleum.

A FEW PETROLEUM PRODUCTS

Lipstick

Sunglasses

Cleats

Ink

Crayons

Roller-skate wheels

Prudhoe Bay, Alaska The Prudhoe Bay oil field in Northern Alaska is the largest oil field in North America. Located on the coast of the Arctic Ocean, the ground is permanently frozen several feet deep in this bitterly cold place. The main Prudhoe Bay oil field has an area of about 1,554 square kilometers (600 square miles). There are smaller oil fields near the main one.

Oil was discovered in the rock under Prudhoe Bay in 1968. Five years later the U.S. government approved plans for a 1,287.5-kilometer (800-mile) pipeline to carry oil from Prudhoe Bay to the port town of Valdez, Alaska. From Valdez, ships carry oil all over the world.

The 19 oil fields in Prudhoe Bay have produced more than 12.8 billion barrels of oil since 1977. Drilling and transporting oil affects the environment. Since 1995, there have been about 400 spills per year from the oil field and pipeline. The total **volume** of oil spilled is about 1.5 million gallons (5,678,118 liters). These spills cause damage to soil and death or injury to wild animals.

The Prudhoe Bay oil field is the largest oil field in North America.

TECHTREK
myNGconnect.com

Prudhoe Bay, Alaska

Digital Library

volume

Volume is the amount of space something takes up.

Natural Gas Natural gas is the third type of fossil fuel found in the earth. It occurs naturally in the gas state. Natural gas also has no color or smell. We often add a smell to natural gas so that we can smell it and avoid being near harmful fumes. Unlike some gases, natural gas burns easily. Natural gas is a very important source of energy throughout the world. We use it for heating homes, cooking, and creating electricity.

Natural gas is made of a mixture of carbon and hydrogen called methane. It also contains small amounts of other substances. Natural gas found in the earth must be refined much like oil before it can be used. Refining natural gas means cleaning out extra substances such as water, oil, gases, sand, and other compounds. After being refined, some of the parts of natural gas are sold separately. Propane (used in grills) and butane (used in lighters) are gases that are refined from natural gas.

The flame of a gas stove is fueled by natural gas.

Drilling for Natural Gas

Geologists locate natural gas in the earth. Once a site has been found, the drilling company works to collect gas. There are two types of drilling used to get natural gas from the earth. One method is to drop a heavy metal bit into the ground over and over. This is called percussion drilling. This process creates a hole from which we can collect natural gas. The other method is called rotary drilling. A drill bit similar to one you may have at home rotates as it digs into the earth.

If the drilling company finds natural gas at a drilling site, workers begin the process of bringing the gas to the surface. First, workers strengthen the wall of the well. Then, they fit pipes and tubes to the top of the well to collect the gas. Sometimes the gas will rise up the tubing on its own. After all, gases expand to fit their containers. In other cases, pumps may need to pull the natural gas from the earth.

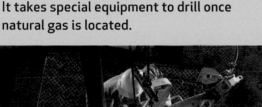
It takes special equipment to drill once natural gas is located.

This series of pipes at the top of a natural gas well controls the flow of matter out of the well.

BECOME AN EXPERT

Natural Gas Under the Sea

At the beginning of 2009, a natural gas deposit was found off the coast of Israel. For a country with few natural energy sources, this was a very big discovery. This was the largest natural gas site ever found in Israel. Where is the site of this natural gas supply? Beneath the ocean floor! After drilling more than 4,572 meters (15,000 feet) into the seafloor, geologists confirmed the gas deposit. The energy company believes that there might be enough natural gas in this site to last 20 years. They plan to start bringing the gas to the surface in 2012.

The Gulf of Venezuela was also the site of a new natural gas discovery in 2009. Venezuela's state oil company believes that this could be one of the largest natural gas deposits in the world. It could contain seven to eight billion cubic feet of gas.

A natural gas deposit was discovered off of the coast of Israel.

One of the largest natural gas deposits in the world was discovered beneath the Gulf of Venezuela.

Energy and the Future

We pump and dig different kinds of matter from the earth to make our energy. But most of this matter is nonrenewable. This means that it takes a long time for the matter to form. Once we use that matter, it's gone. At least, it's gone for millions of years! What other kinds of matter can we use for energy?

One type of matter we can use for energy is the air. Wind power can create electricity. Wind is simply the movement of air. Windmills capture the energy of wind and turn it into electricity. Wind energy is a renewable resource.

Water is a liquid source of power. Water that flows swiftly has a great deal of energy. The water pushes against turbines. The moving turbines can create electricity from the energy of the moving water.

Earth is rich with resources, matter that we use every day. Taking care of Earth and its resources is one way to preserve that matter for use in the future.

CHAPTER 1

SHARE AND COMPARE

Turn and Talk How does the matter that we can find in the Earth affect our daily lives? Form a complete answer to this question along with a partner.

Read Select two pages in this section. Practice reading the pages. Then read them aloud to a partner. Talk about why the pages are interesting.

my **SCIENCE** notebook

Write Write a conclusion that tells the important ideas you learned about natural resources. State what you think is the Big Idea of this section. Share what you wrote with a classmate. Compare your conclusions. Did your classmate come to the conclusion that natural resources exist in all three states of matter?

my **SCIENCE** notebook

Draw Draw a picture that represents the origin of a fuel that comes from the earth. Add a caption to your drawing. Then compare your drawings to those of other classmates. Create a display with all your drawings to show the various ways that fuels come from Earth.

CHAPTER
2

HOW CAN MATTER CHANGE?

Everything around you is made of matter. The glacier in this picture is made of matter, too. Matter does not always stay the same. Huge pieces of this glacier in Argentina are crashing down in to the water below. The matter in this glacier is changing shape and size.

TECHTREK
myNGconnect.com

The falling pieces of ice are made of the same kind of matter as the large glacier. The glacier is changing in shape and size.

54   55

After reading Chapter 2, you will be able to:

• Understand that an object undergoing a physical change keeps its chemical composition.
PHYSICAL CHANGES

• Understand that the mass of water remains constant as it changes state.
PHYSICAL CHANGES

• Name and describe various physical changes, such as tearing, crushing, freezing, melting, boiling, evaporation, and condensation. **PHYSICAL CHANGES**

• Recognize that matter has observable physical properties. **PHYSICAL CHANGES**

• Understand that freezing, melting, condensation, evaporation, and boiling are changes in state that are caused by changes in temperature. **PHYSICAL CHANGES**

• Recognize that, during a change of state, heat energy is absorbed or released.
PHYSICAL CHANGES

• Describe characteristic physical properties such as boiling point, melting point, and solubility. **PHYSICAL CHANGES**

• Recognize evidence of a chemical change and understand that a chemical change can create new substances with different properties. **CHEMICAL CHANGES**

• Science in a Snap! Name and describe various physical changes, such as tearing, crushing, freezing, melting, boiling, evaporation, and condensation. **PHYSICAL CHANGES**

CHAPTER

2

HOW CAN MATT

Everything around you is made of matter. The glacier in this picture is made of matter, too. Matter does not always stay the same. Huge pieces of this glacier in Argentina are crashing down in to the water below. The matter in this glacier is changing shape and size.

ER CHANGE?

The falling pieces of ice are made of the same kind of matter as the large glacier. The glacier is changing in shape and size.

SCIENCE VOCABULARY

physical change
(FI-si-kul chānj)

A **physical change** is when matter changes to look different but does not become a new kind of matter. (p. 59)

Crushing this can changes its size and shape, but the can is still made of aluminum.

evaporation
(ē-va-pōr-Ā-shun)

Evaporation is the physical change of matter from a liquid state to a gaseous state. (p. 64)

Lake Powell, Utah: The different colored bands on the rocks show where evaporation has caused the water level to drop over the years.

my Science Vocabulary

chemical change
(KEM-i-kul chānj)

condensation
(kon-din-Sā-shun)

evaporation
(ē-va-pōr-ā-shun)

physical change
(FI-si-kul chānj)

TECHTREK
myNGconnect.com

Vocabulary
Games

condensation
(kon-din-SĀ-shun)

Condensation is the physical change of matter from a gaseous state to a liquid state. (p. 66)

As the warm air touches this cold glass of lemonade, water vapor changes to liquid water, forming condensation on the outside of the glass.

chemical change
(KEM-i-kul chānj)

A **chemical change** is a change in matter that forms a new substance with different properties. (p. 68)

Oxygen reacts with the iron in these cans to cause a chemical change. The result is rust.

Physical Changes

Look around you. What can you see? Matter! Anything that has mass and takes up space is matter. Matter takes different forms—solids, liquids, and gases.

If you've seen ice melt, or mixed paint to form a new color, then you already know that matter can change. Think of a block of ice. Ice is solid water. If a sculptor carves a block of ice, the ice has different properties. Its size and shape change. The matter itself, however, hasn't changed. It's still ice. What would happen to the amount of matter if the ice melted? The amount of liquid matter would be the same as in the ice.

This ice sculpture in Harbin, China, shows that the shape of ice can be changed in magnificent ways.

Imagine dyeing white t-shirts different colors. You'd be changing the color of each t-shirt. Color is a property of matter. A property is something about an object that you can observe with your senses.

Other physical properties include texture, shape, size, odor, volume, weight, mass, and temperature. A physical change is a change in the physical properties of matter. Dyeing a t-shirt creates a physical change.

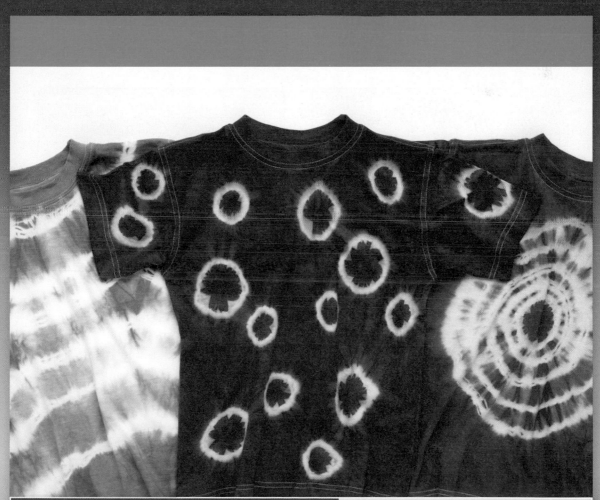

These t-shirts used to be white, but now their colors have changed. Changing color is a physical change. Even with the new colors, the t-shirts are still made of the same matter.

You can see physical changes every day. Cooking involves many physical changes. Tearing lettuce, crushing walnuts, and melting butter are all physical changes.

Imagine stepping on a can and crushing it. You would change some properties of the can. The can would have a different shape. It might seem to be a different size. It would still be a can, though. The can would still be made of solid metal. Look at the table on the next page to find out about different types of physical changes.

Science in a Snap! Time for a Change

Take one of the items in front of you and describe its physical properties.

Make a physical change—tear, mold, fold, crush.

Have a partner identify what kind of physical change was made.

EXAMPLES OF **PHYSICAL CHANGES**

Read the information in the table to learn about some different physical changes.

TECHTREK
myNGconnect.com

Enrichment
Activities

TEARING
If you tear something, you pull it into pieces.

CRUSHING
If you crush something, you squeeze or press it so that its shape changes or it breaks.

FREEZING
When liquid freezes, it becomes solid.

MELTING
Some solids, such as ice, melt when the temperature rises to a certain point.

BOILING
When it reaches a certain temperature, a liquid begins to bubble and change into gas.

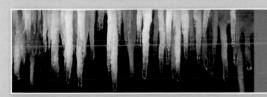

EVAPORATION
Evaporation occurs when the temperature of a liquid is raised enough to turn it into a gas.

CONDENSATION
Condensation occurs when warm water vapor is cooled and becomes a liquid.

You're already familiar with one of the most important substances on Earth—water. Water can be found on Earth as a solid, a liquid, and a gas. The change from liquid to solid, or freezing, is a physical change. Water is the same substance whether it is solid or liquid.

What causes the physical changes in water? Physical changes in water are caused by temperature changes. Other kinds of matter undergo physical changes because of temperature, too. Wax is matter that can change from solid to liquid when the temperature rises. Even glass and metals, heated to very high temperatures, can melt, changing to liquids.

Water freezes on branches when the temperature drops below 0° Celsius (32° Fahrenheit).

Freezing and Melting Picture a stream. The liquid water in a stream flows for most of the year. In winter, if the temperature is low enough, the water's state changes. The water becomes solid ice. The change from a liquid to a solid is called freezing. Many liquids freeze. Water freezes at 0° Celsius (32° Fahrenheit).

The temperature at which a liquid freezes is called its freezing point. Freezing causes the atoms in a liquid to move so slowly that the liquid becomes a solid. 0°C (32°F) is also the melting point of ice. If you raise the temperature, ice undergoes a physical change, melting to become liquid water.

When the temperature reaches the melting point, solid ice begins to melt, forming water.

Evaporation In the small picture, water is boiling in a pot. What do you notice about the water? Bubbles are forming. Those bubbles are the gas form of water—water vapor. Water begins the physical change of turning from liquid to gas at 100°C (212°F). The temperature at which matter changes from liquid to gas is the matter's boiling point. Other liquids besides water can undergo a physical change at their boiling points.

The physical change of matter from liquid to gas is called **evaporation** . When water evaporates, it becomes water vapor. Water vapor is invisible. Then why do you see steam rising over a pot of boiling water? Steam happens when invisible water vapor enters the cooler air. The cooler air makes the gas turn back into a liquid.

At its boiling point, water begins to change from a liquid to a gas.

Water does not need to boil to evaporate. It can evaporate at lower temperatures, too. In nature, energy from the sun and wind can cause water to evaporate. As the sun warms the water in lakes, ponds, rivers, and even the huge oceans, the top of the water can become water vapor, rising into the air.

This is a much slower process than boiling. If you observe carefully, you can see the effects of evaporation. Look at the photo below. The drop in the water level happened because of evaporation over many years.

TECHTREK
myNGconnect.com

Digital Library

The bands on the rocks show that the water level in Lake Powell in Utah has decreased over the years.

Condensation If you have ever seen clouds, dew, or fog, you have seen **condensation**. Condensation is the opposite of evaporation. It is the physical change of matter from a gas to a liquid.

Have you ever wondered why your glass of ice water may get beads of water on the outside? The warm water vapor in the air touches the cold glass. The coldness of the glass causes the temperature of the water vapor to drop. The water vapor changes from a gas to a liquid on the side of your glass. The same thing happens when warm daytime air is cooled at night. The condensation you see and feel on the grass is called dew.

The beads of water on this glass are formed by condensation. Condensation is a physical change of water from the gaseous state to the liquid state.

Water vapor condensed to form dew droplets on this spider web.

Why is condensation important? The air is filled with invisible water vapor. Particles of water vapor cling to dust particles. As the water vapor rises, the air cools, making the vapor condense on the dust. These droplets combine with other droplets to form clouds.

When the droplets become too heavy, they fall back to Earth. The water from clouds allows living things to grow and gives us water to drink. If condensation didn't form clouds, we would not be able to live on Earth.

TECHTREK
myNGconnect.com

Clouds release water that is essential for life on Earth.

Digital Library

Before You Move On

1. What is the difference between evaporation and condensation?
2. Explain why beads of water might form on the outside of a glass holding a cold beverage but would not form on the outside of a mug containing a warm liquid.
3. **Infer** How does condensation support life on Earth?

Chemical Changes

If you rip a piece of paper into pieces, you have caused a physical change. Each tiny piece of paper is still a piece of paper. What if you burned the piece of paper instead? The paper would change into a new substance and could not turn back into paper again. A **chemical change** is a change in matter in which new matter forms. The new matter has different properties. Chemical changes are permanent. Once matter goes through a chemical change, it does not change back to its original form.

How can you tell that a chemical change has occurred? You will see one or more of the following signs:

✓ Gas forms.
✓ Heat is given off or absorbed.
✓ A solid forms or disappears.
✓ A new color occurs.
✓ A new odor is created.
✓ Light is produced.

A chemical change in a lightning bug's body causes the light.

During the restoration of the Statue of Liberty, some of the copper covering was replaced with new copper.

The designer of the Statue of Liberty, French sculptor Frederic Auguste Bartholdi, chose an excellent covering for this symbol of freedom. The statue is covered in copper, which does not rust. Instead, the copper reacts with the oxygen in the air.

The result is a blue-green coating that forms on the copper. This coating has protected the skin of Lady Liberty for over 100 years. It was one of the few parts of the statue that did not need extensive repair when it was renovated for its 100th birthday.

Burning You read about the chemical change on the surface of the Statue of Liberty. The covering of the statue did not turn green quickly. The change took a long time. Some other chemical changes, like rust, also happen very slowly.

Other chemical changes, however, can happen very quickly. Fire can happen when oxygen in the atmosphere reacts with fuel, like wood. To make fire, heat needs to be added to the wood.

Heat can come from a match or even from a lightning strike. Once the wood reaches about 150°C (300°F), fire occurs. The wood forms new matter. Burning wood forms gases and tiny solid particles that you can see as smoke.

Burning wood creates a chemical change that is easy to see.

Burning the wood causes chemical changes, not physical changes. How can you tell? Look at the signs listed in the chart. Gas forms (in the smoke); heat is given off; a solid (wood) disappears, and a new solid (ash) forms; and light is produced. These changes are permanent. No one can ever change ash back into wood.

SIGNS OF A CHEMICAL CHANGE

✓ Gas forms.
✓ Heat is given off or absorbed.
✓ A solid forms or disappears.
✓ Light is produced.

Wood

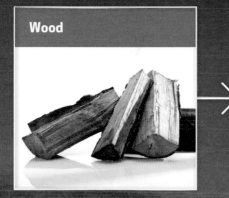

Burning wood

Ash

Rusting Picture an old hubcap on a car that has been left outside for a long time. The hubcap used to be shiny, but now it is covered with a dull reddish-brown substance. The hubcap is made from steel.

Steel is a metal that includes iron. Steel is a strong metal, used for everyday objects from surgical needs to washing machine parts. But steel has one problem. It can rust.

Rust needs both iron and oxygen to form. Rust's chemical name is iron oxide.

Rusting is a chemical change over time. Iron, water, and oxygen from the air are needed for rust to form.

How does rust affect iron? Rust can create holes in pipes, leading to water or gas leaks. Rust can also create holes in the bodies and engine parts of cars, making the cars unsafe to drive.

Digital Library

TECHTREK
myNGconnect.com

The rust on these trucks formed when iron reacted with water and with the oxygen in the air.

Cooking You already know that cooking involves physical changes. But cooking involves many chemical changes, too. Chefs use both temperature and special ingredients to make chemical changes happen in cooking.

Which signs of chemical change can you observe when an egg cooks? The egg becomes hotter. The clear egg liquid turns into a white solid. A new odor is produced. Raw eggs have little odor. Cooked eggs have a distinctive smell. A cooked egg cannot be turned back into a raw egg.

Cooking an egg causes a chemical change in the egg.

Souring If you have ever kept milk long past its expiration date, you may have observed a chemical change. Milk has bacteria in it. The amount of bacteria in fresh milk is harmless to people. Over time the bacteria start to feed on the sugar in milk. As the bacteria grow, chemical changes happen in milk.

What signs of chemical change can you observe? Sour milk has a different taste than fresh milk. It also has a strong odor. A solid forms in sour milk, too. If you take a look at the milk, you can see clumps of matter called curds.

fresh milk

sour milk

Before You Move On

1. What are some common chemical changes? Name at least two.
2. How does oxygen play a part in chemical changes?
3. **Analyze** When ice melts, a solid disappears. Why isn't ice melting a chemical change?

WILDFIRES
IN THE UNITED STATES

A firefighter supervisor drives through flames that have jumped the road in Seeley Lake, Montana.

National Geographic photographer Mark Thiessen captured the image above while traveling with a firefighter supervisor through a forest fire in Seeley Lake, Montana. More than 64,000 fires rage throughout the United States each year, destroying around six million acres of land. The "fire season" lasts from late spring into the fall. How do the fires start? How do they stay burning? And how do they change the matter around them?

Lightning can start wildfires. People can start fires, too. People can be careless, leaving fires unattended. These fires can spread and burn out of control. Once a fire starts, the weather affects it. If temperatures are high, winds are strong, and the air is dry, fires can burn longer and spread more quickly. Under these conditions, fires are hard to get under control.

Only three things are needed for a fire: fuel, heat, and oxygen. Trees, bushes, twigs, sticks, grasses, and other plant life are all sources of fuel for a wildfire. Heat can be created in several ways: lightning striking a tree, a hot match thrown away carelessly, or the remains of a campfire. Oxygen, of course, is in the air all around.

heat

oxygen   fuel

If one of these three things is taken away, fire cannot burn.

Under dry conditions, fire can spread quickly and be hard to put out.

Fire needs fuel and oxygen to keep going. Trees and vegetation on the floor of a forest become fuel for the fire. As long as the fire has oxygen and fuel, it will continue to burn.

To put out a large fire, firefighters must remove at least one of the things fire needs: fuel, heat, or oxygen. One way that firefighters try to contain a wildfire is by taking away its fuel supply. To do this, they might carefully burn down a section of vegetation that is in the path of the fire. This removes the fuel supply from the fire.

Firefighters in Seeley Lake use flares to start a fire. This fire will take away some of the fuel in the path of a raging wildfire.

Another way to control a fire is to take away the oxygen supply. Fire retardant foam is made of materials that do not burn. Firefighters drop this foam from airplanes. The foam keeps oxygen away from the fire. Finally, firefighters try to reduce the heat produced by fire by spraying water on it.

Fire is not all bad. It clears away dead vegetation. It also creates a lot of ash. Ash has nutrients in it. These nutrients feed and support the growth of new plants.

Firefighters remove the fuel, heat, or oxygen so they can put out the fire.

Conclusion

Everything that has mass and takes up space is matter.
Matter can undergo physical and chemical changes.
Physical changes, such as crushing, tearing, melting, and
freezing, change properties of matter. The kind of matter,
though, does not change. Water can undergo many physical
changes, such as freezing, melting, boiling, evaporation,
and condensation. Chemical changes, such as rusting,
burning, and souring, cause new substances to form.
The changes are permanent.

Big Idea Matter can change in different ways: physically
and chemically.

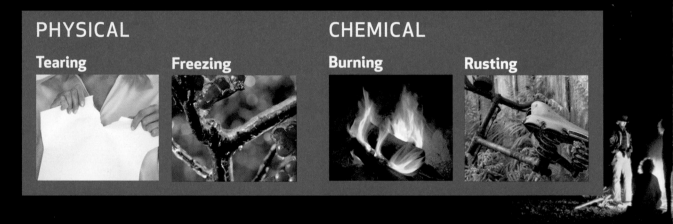

PHYSICAL

Tearing

Freezing

CHEMICAL

Burning

Rusting

Vocabulary Review

Match the following terms with the correct definition.

A. **physical change**

B. **chemical change**

C. **evaporation**

D. **condensation**

1. The physical change of matter from the liquid state to the gaseous state

2. A change in which matter changes to look different but does not become a new kind of matter

3. The physical change of matter from a gaseous state to a liquid state

4. A change in matter that forms a new substance with different properties

Big Idea Review

1. **List** List three types of physical changes that a substance can undergo.

2. **Restate** State in your own words what happens during a chemical change.

3. **Explain** If a sculptor carves a block of marble into a sculpture, what kind of change has taken place in the marble?

4. **Classify** Give one example of a physical or chemical change that is beneficial for human beings. Give one example of a physical or chemical change that can be harmful for humans.

5. **Infer** Suppose you found a piece of unknown metal outside. What might be one sign that this metal contains iron?

6. **Analyze** A friend tells you that burning a wax candle causes both a physical and chemical change. Is your friend correct? Explain why or why not.

my SCIENCE notebook

Write About How Matter Changes

Explain What is happening in this photo? Is this an example of a physical change or a chemical change? How do you know?

PHYSICAL SCIENCE EXPERT: CUTTING-EDGE CHEF

Chef Wylie Dufresne

Wylie Dufresne (pronounced Du-frane) is a world-famous chef with his own restaurant in New York City. His use of high-tech equipment and scientific experimentation in his kitchen allows him to make delicious and interesting foods like peanut butter pasta and freeze-dried polenta.

Wylie Dufresne carefully studies physical and chemical changes in ingredients to create some amazing dishes!

NG Science: Who are you and what does someone like you do?

Wylie Dufresne: I'm a chef. I own my own restaurant. A chef designs a menu. Chefs are in charge of getting all the ingredients. A chef and his or her staff prepare the ingredients and cook food.

NG Science: What is the coolest part of your job?

Wylie Dufresne: Eating is pretty cool! I like working together with my staff. And I love the kitchen in my restaurant. It's a great space to work.

NG Science: What's a typical day for you?

Wylie Dufresne: We start in the morning and we're still cooking at 1:00 a.m. It's exciting when we get busy. It's a lot like playing a sport!

NG Science: Did you always want to be a chef?

Wylie Dufresne: No. I didn't start cooking until I was in college. I worked at a restaurant making pizzas one summer. I started reading cookbooks and trying to cook new things. I tried to understand the chemistry of what happens to food when you cook it.

TECHTREK
myNGconnect.com

Student
eEdition

Digital
Library

NG Science: What kind of contribution do you hope to make through your work?

Wylie Dufresne: I hope that our work at the restaurant will add to the understanding of cooking. It will make a contribution to what we know about food and how to prepare it.

NG Science: Do you see a strong connection between what you do and Physical Science?

Wylie Dufresne: Yes. Cooking is chemistry. Chefs need to understand the chemical changes that happen during cooking. Chefs need to know how and why food behaves the way it does when you cook.

TECHTREK
myNGconnect.com

Digital
Library

Wylie Dufresne is heating food in a pan. Do you think he is causing physical changes, chemical changes, or both?

Physical and Chemical Changes:
In the Culinary Arts

Culinary Chemistry Chefs are a lot like scientists. They use chemistry every day in their jobs.

All around the kitchen, **physical** and **chemical changes** take place. Take a tour of a kitchen. You might start with the preparation area. This is where many physical changes happen. Chefs cut up the vegetables that they will use that day. They have to be careful because they use very sharp knives. This is also the place where seasonings are mixed. Seasonings are used to bring out the flavors in cooked food.

Adding seasoning to food can cause a chemical change. You can't remove the seasoning to change the food back.

Slicing tomatoes causes a physical change.

physical change

A **physical change** is when matter changes to look different but does not become a new kind of matter.

chemical change

A **chemical change** is change in matter that forms a new substance with different properties.

Chopping food is a physical change.

Coolers keep vegetables and plates cool at the salad station. The next time you order a salad, look at your plate. Do you see **condensation** ? That condensation may be the result of bringing the plate out of the freezer and onto your table.

condensation

Condensation is the physical change of matter from a gaseous state to a liquid state.

85

BECOME AN EXPERT

Sautéing, grilling, baking, and roasting all cause chemical changes in food. Once a piece of meat is grilled, the changes in the meat are permanent. The meat cannot return to its original form.

Chefs know how to use physical changes in water to cause chemical changes in food. Boiling water can create steam. Chefs use steam to cook vegetables. The vegetables undergo chemical changes. They cannot return to their original forms, even when cooled.

TECHTREK
myNGconnect.com

Digital Library

Sautéing means to cook food quickly at a very high temperature.

Steaming is a healthy way to cook foods such as fish.

Baking creates chemical changes, too. When you add yeast to other ingredients to make bread, the yeast reacts with the sugar. A gas called carbon dioxide forms. The gas makes the bread rise, so it becomes light and fluffy.

People of many cultures around the world bake breads without using yeast. Without a chemical change to create carbon dioxide, the breads don't rise. Matzo and tortilla are a few examples of flatbreads from around the world.

Kneading dough causes the yeast to react with other ingredients. The yeast causes a chemical change in the bread.

Yeast causes bread to rise. Which breads were not baked with yeast?

Food Preservation For thousands of years, people have been finding ways to preserve food. *Preserve* means "to make something last." There are many ways food can be preserved. Preserving food can involve chemical or physical changes.

Salting and pickling are two of the oldest ways to preserve foods. Before there were refrigerators, people would pack their meats tightly in salt. The salt would pull all of the water from the meat. Without water, the bacteria could not grow, and the meat would not spoil. In pickling, salt and vinegar are used to preserve meats, fruits, and vegetables. The vinegar keeps the bacteria from growing. The chemical changes of salting and pickling can make some foods last for years.

Today, many types of meat are still salted.

Today, pickling is used to make a variety of foods. Cucumbers can be changed to pickles that are sour or sweet.

Another way to preserve food is by using the physical change of evaporation to remove the water from foods. There are many examples of dried foods. Dried pastas and powdered milk are preserved by **evaporation.** When the water is removed from foods, bacteria cannot grow. This makes foods last much longer.

APRICOTS: FRESH TO PRESERVED

Fresh apricots are very moist.

Dried apricots last much longer than fresh ones.

evaporation
Evaporation is the physical change of matter from a liquid state to a gaseous state.

What do astronauts eat in space? Freeze-dried foods are the best choice. They often take up less space than fresh foods. They can sometimes be eaten dried. Freeze-drying is a physical change. The food has water removed from it, but the make-up of the food is the same.

Astronauts eat three meals a day in space. These meals are carefully chosen to make sure the astronauts receive all of the vitamins and nutrients they need. Some foods, such as fruit, can be eaten in their natural form. Other foods, such as spaghetti or macaroni and cheese, are freeze-dried. The astronauts must add water to these foods before they can be eaten.

These foods were prepared for space travel.

Eating in space has its challenges. With no gravity, food can just float away! Like many people, astronauts like to add salt and pepper to their food. However, salt and pepper could cause problems. In space, salt and pepper would float away. It could damage equipment. Astronauts use special salt and pepper in liquid form. Crumbs from bread can also cause problems in space. This is why astronauts eat tortillas instead of bread. Tortillas don't crumble like bread does.

Tortillas are safer than slices of bread for astronauts in space.

Food for space travel is designed to stay in place.

CHAPTER 2

SHARE AND COMPARE

Turn and Talk How are physical and chemical changes both important to cooking? Form a complete answer to this question with a partner.

Read Select two pages in this section. Practice reading the pages. Then read them aloud to a partner. Talk about why the pages are interesting.

Write Write a conclusion that tells the important ideas you learned about chemical and physical changes in cooking. State what you think is the Big Idea of this section. Share what you wrote with a classmate. Compare your conclusions. Did your classmate understand that both chemical and physical changes are important in cooking?

Draw Draw a picture that represents a physical change or a chemical change in cooking and add a caption that explains the change. Combine your drawing with those of other classmates to make a Physical and Chemical Changes Cookbook.

HOW DO YOU DESCRIBE
FORCE
AND THE LAWS OF MOTION?

Things are in motion all around you. Every motion that you see is the result of forces. At the circus, you may see people flying through the air during a trapeze act. The acrobats swing, hold, and twist. They use force in each move to perform the routine.

TECHTREK
myNGconnect.com

The acrobat uses force to move.

After reading Chapter 3, you will be able to:

- Describe the motion of an object in terms of position, direction, distance, and time.
  FORCE AND MOTION

- Recognize that a force is a push or a pull, and can he contact (i.e., friction) or non-contact (i.e., gravity). **FORCE AND MOTION, FRICTION AND AIR RESISTANCE, GRAVITY**

- Identify the forces applied to an object as balanced or unbalanced.
  NEWTON'S LAWS OF MOTION

- Recognize that an object will stay at rest or continue at a constant velocity unless acted on by an unbalanced force. **NEWTON'S LAWS OF MOTION**

- Recognize that when force is applied to an object and it does not move, it is the result of balanced forces being applied to the object. **NEWTON'S LAWS OF MOTION**

- Recognize that for every force applied to an object, there is an equal, but opposite, force applied by the object. **NEWTON'S LAWS OF MOTION**

- Science in a Snap! Recognize that for every force applied to an object, there is an equal, but opposite, force applied by the object. **NEWTON'S LAWS OF MOTION**

HOW DO YOU DESCRIBE FORCE AND THE

Things are in motion all around you. Every motion that you see is the result of forces. At the circus, you may see people flying through the air during a trapeze act. The acrobats swing, hold, and twist. They use force in each move to perform the routine.

TECHTREK
myNGconnect.com

Student
eEdition

Vocabulary
Games

Digital
Library

+
Enrichment
Activities

The acrobat uses force to move.

LAWS OF MOTION?

SCIENCE VOCABULARY

force (FORS)

A **force** is a push or a pull. (p. 98)

Forces acting on a object can cause the object to move.

motion (MŌ-shun)

Motion is a change in position. (p. 98)

The ball and some of the cans are in motion.

my Science Vocabulary

force (FORS)   **motion** (MŌ-shun)

gravity (GRA-vi-tē)

TECHTREK
myNGconnect.com

Vocabulary Games

gravity (GRA-vi-tē)

Earth's **gravity** is a force that pulls things to the center of Earth. (p. 106)

Gravity pulls the girl into the water.

Force and Motion

Look at the children below. They are taking their dog to the park. They are pulling him in a wagon. The children are using a force to make the wagon move. Force is a push or a pull that causes an object to move.

The children are using a contact force to make the wagon move. To use contact forces, the children have to touch the wagon when they move it.

Think about the wagon that the children pulled. Before they pulled the wagon, it was still. When they used force, the wagon moved to a new position. The wagon was in motion. Motion is a change in position.

The children are pulling the wagon. They are using force to pull the dog.

Direction Direction is one way to describe motion. You can use compass directions, such as north and south. Or you can use many other directions. The bird flew up. The car turned left. Objects can also move toward and away. The rabbit ran toward the gate. The dog ran away from the house. Direction helps describe where an object is going.

Direction also helps to describe forces as well. In the picture, the children are pulling the wagon. The wagon is moving toward the children. So, the direction of the force being exerted on the wagon is toward the children.

99

Speed Telling how fast an object is moving is another way to describe the object's motion. Speed tells you how fast an object is moving. To find the speed of an object, you need to know two things. First, how far did it go? Second, how long did it take to go that far? In other words, you need to know distance and time to find speed.

The speed of an object can be very slow. For example, earthworms and snails move slowly. But the speeds of other things can be very fast. A jet plane can fly at 880 kilometers per hour (547 miles per hour).

The ostrich is the fastest animal on two feet. An ostrich can reach speeds of 70 kilometers per hour (43 miles per hour).

THE SPEED OF ANY MOVING OBJECT CAN BE FOUND USING AN EQUATION.

$$speed = \frac{distance}{time}$$

For example, an ostrich can run 30 meters (about 98 feet) in 2 seconds. So the ostrich's speed is:

$$speed = \frac{30 \text{ meters}}{2 \text{ seconds}}$$

$$speed = 15 \text{ meters/second}$$

OSTRICH **SPEED**

TIME (SECONDS)	DISTANCE (METERS)
1	15
2	30
3	45
4	60
5	75
6	90

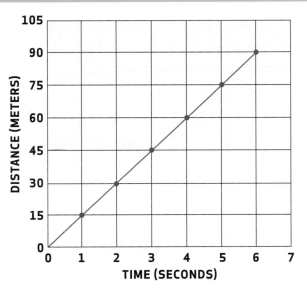

Speed can be shown on a graph.

1. Put time on the x-axis. Put distance on the y-axis.

2. Using information from the chart on the left, make a dot on the graph to show where the ostrich is at each point in time.

3. Connect the dots with a line.

Before You Move On

1. What is a force?
2. How are force and motion related?
3. **Analyze** How can you find the speed of an object in motion?

Friction and Air Resistance

Forces can make things move faster. If you kick a soccer ball, it moves quickly down the field. But forces can also make things move slower and even stop. One force that slows things down is friction. Friction is a force that acts when two surfaces are touching. Friction is a contact force that can make moving objects slow down or stop. Think about that soccer ball you just kicked. It will slow down and finally stop. Why? Because of friction between the ball and the ground.

The amount of friction between two surfaces depends on the texture of the surfaces that are touching. Rough surfaces usually cause more friction than smooth surfaces. The grass in the outfield of a baseball diamond is rougher than the infield dirt. So there will be more friction when a ball rolls on the grass than there will be when the ball rolls on the dirt.

Friction between the baseball player and the ground slows the player down as he reaches a base. If there were no friction, he would go sliding past the base and could be tagged out!

Suppose that you have a ball rolling on the grass and a second ball rolling on the dirt at the same speed. The ball on the grass will slow down more quickly and will stop sooner because of the greater friction. Friction is needed for almost everything that you do. Friction between your feet and the ground lets you walk. Friction between your hand and your pencil lets you hold the pencil. Friction between the pencil and the paper lets the pencil leave a mark.

Air Resistance Another kind of friction is air resistance. Air resistance works against the motion of objects that are traveling through the air. Air is all around, so air resistance is always acting on moving things.

Air resistance depends partly on the speed of the object moving through the air. If you run instead of walk, you will feel more air resistance against your body.

The force of air resistance pushes up on the parachute. This force causes the parachute and the person attached to it to slow down.

Air resistance also partly depends on the shape of the object traveling through air. Larger or wider things usually feel more air resistance. A large van will have more air resistance than a small sports car even if the two cars are moving at the same speed.

Like other kinds of friction, air resistance can be very useful. Skydivers take advantage of air resistance every time they jump out of airplanes. As a skydiver falls through the air, her body picks up speed. Well before a skydiver reaches the ground, she will open her parachute. The parachute is very wide and it increases the air resistance of the falling skydiver. The air resistance on the parachute slows the skydiver down so much that she can land on the ground safely.

Before You Move On

1. What is friction?
2. Why is it easier to slide on ice than on a concrete sidewalk?
3. **Apply** A crumpled sheet of paper and a flat sheet of paper are dropped from the same height. Why does the crumpled sheet hit the ground faster?

Gravity

Suppose you drop a ball. The force that causes the ball to fall to the ground is **gravity** . The gravity between Earth and the ball causes the ball to move to the center of Earth.

This happens even though Earth and the ball are not touching each other. Gravity is a non-contact force. Gravity pulls objects downward unless something holds them up.

Although the Earth is not touching this swimmer, she is pulled toward Earth by Earth's gravity.

Look at the four photos of the girl. She is jumping into a lake. Have you ever jumped into the water? You may stand on a platform near the edge of the water. Then you jump! What happens? You land in the water and make a big splash.

Why did this happen? Earth's gravity is pulling you to the center of Earth. When you jump off the platform, Earth's gravity pulls you downward into the water. Earth's gravity is a force that pulls on everything on Earth.

Before You Move On

1. Why is gravity considered a non-contact force?
2. What does gravity do to objects on Earth?
3. **Predict** Imagine kicking a ball in the air. What does the ball do? Why?

NATIONAL GEOGRAPHIC

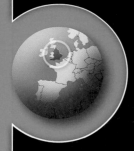

SIR ISAAC NEWTON
GRAVITY AND
THE LAWS OF MOTION

Isaac Newton was one of the greatest scientists of all time. He was born in 1642 in England. When he was 16, his mother took him out of school to begin working on the farm. However, Newton did not like being a farmer. And so, when he was 18, he went to college.

Newton studied math, physics, and astronomy. After he graduated, he stayed to work as a teacher. But later that year, the Great Plague broke out, and the school was closed. Newton had to go back home to the family farm.

Isaac Newton made many important contributions to science. Not only did he describe the laws of motion and of gravity, he also built a telescope using mirrors, developed a new kind of math, and studied light and astronomy.

After he returned home, Newton continued learning on his own. Some of Newton's greatest discoveries happened during this time. One such discovery occurred while observing the effects of gravity.

Newton realized that gravity is a force. It caused objects to fall to Earth. He realized that the same force must also keep the moon in orbit. Newton discovered that gravity depends on the mass of objects and the distance between them. He worked out an equation to find the strength of gravity. Scientists still use his equation today!

Newton did not publish his ideas about gravity until 1687. These ideas were published in three books. Together, they are called *Principia*. The *Principia* includes some of the most important ideas ever written about science.

MATHEMATICAL

PRINCIPLES

OF

NATURAL PHILOSOPHY.

By Sir ISAAC NEWTON, Knight.

TRANSLATED INTO ENGLISH, AND ILLUSTRATED WITH A
COMMENTARY,

By ROBERT THORP, M.A.

VOLUME THE FIRST.

LONDON:
PRINTED FOR W. STRAHAN; AND T. CADELL, IN THE STRAND.
MDCCLXXVII.

The first book of *Principia* begins with a description of three laws of motion. These three laws are now called Newton's laws of motion. The laws describe how objects move when acted on by forces. Each law is listed below.

The third book of *Principia* describes how mass and distance affect the gravity between two objects. The third book also explains how gravity affects the motion of the planets and other bodies in the solar system.

By writing the *Principia*, and making many other important discoveries, Newton made many important contributions to science.

Newton discovered that sunlight could be separated into colors.

FIRST LAW OF MOTION
An object at rest stays at rest and an object in motion stays in motion at the same speed and in the same direction unless acted on by an unequal force.

SECOND LAW OF MOTION
The speed of an object depends on the mass of the object and the size of the force acting on it.

THIRD LAW OF MOTION
For every action, there is an equal and opposite reaction.

Newton's Laws of Motion

Look at the girls on the slide. Many different forces are acting on the girls. The water is pushing them down the slide. Gravity is pulling them down the slide. These forces are unbalanced forces. Unbalanced forces are not equal. The forces that are moving the girls down the slide are greater than the forces that are holding them up. Because of these unbalanced forces, the girls slide down into the pool.

Sometimes different forces cancel each other out. Suppose you and your friend are both pulling on a book. You notice that even though you and your friend are pulling really hard, the book doesn't really seem to change position. This is because the forces that you and your friend are applying to the book are the same and cancel each other out. These forces are now balanced, or equal. Unlike the unbalanced forces that are pulling the girls down the slide, the forces that you and your friend are exerting on the book are balanced, and so the book does not move.

TECHTREK
myNGconnect.com

Enrichment Activities

The force of gravity pulls the girls down the slide.

1st Law

Newton's First Law Any object that is not moving is said to be at rest. Many of the cans in the photo are at rest. According to Newton's first law, an object will stay at rest unless acted on by a force. In other words, something that is not moving will not start moving until it is pushed or pulled. When the cans are hit by the ball or other falling cans, they will move.

The law also states that an object will continue in the direction and at the speed that it is moving unless acted on by a force. After the ball hits the cans, it may hit a wall and then stop. The wall exerts a greater force on the ball than the ball exerts on the wall, and so the ball stops moving.

These cans will stay at rest unless they are acted upon by a force, such as the moving ball.

Newton's Second Law Look at the photos on these pages. Which person is using more force? How can you tell?

You know that Newton's first law of motion tells you that a push or a pull can change the motion of an object. His second law of motion tells you that a harder push or pull will cause a greater change in the object's motion.

It also tells you that an object with more mass is harder to move than an object with less mass.

In the first photo, the big splash shows a greater change in the water's motion. It shows a person using a lot of force to make a large amount of water move quickly into the air. The small splash in the second photo, shows less force being used.

2nd Law

Another important part of Newton's second law is that the speed at which an object moves depends on the mass of an object and the size of the force acting on it. Suppose you were throwing a baseball. The force you exert, or the speed at which you move your arm, affects the speed at which the baseball moves.

However, if you tried to throw a bowling ball, you would need to a use a lot more force than when throwing the baseball. As the object's mass increases, so does the amount of force needed to move the object at the same speed. Since the bowling ball has more mass than the baseball, it is harder to throw at the same speed.

A small splash shows that less force is being used because a smaller amount of water is being moved.

Newton's Third Law Look at the photo below. How does the woman get the kayak to move? Newton's third law of motion helps explain how she does it.

When she paddles the kayak, she pushes the paddle against the water. The water pushes back against the paddle. The two forces exert in opposite directions. The result of the forces is that the kayak moves forward.

Newton's third law of motion is often summarized as, "For every action, there is an equal and opposite reaction." When the woman forces the paddle through the water, which is her action, the water pushes back against the paddle with an equal and opposite reaction, and off she goes!

3rd Law

Sometimes the equal and opposite forces cause no change in motion. For example, the people that are sitting at the beach are exerting force on their chairs. The chairs are exerting an equal and opposite force on the people. The result of these forces is that the people are seated in their chairs and do not fall down. The forces being applied are balanced.

When you are sitting, you push down on the chair. The chair pushes up on you.

Science in a Snap! Balloon Rocket

Blow up a balloon. Tape a straw on top of the balloon while you hold the end of the balloon.

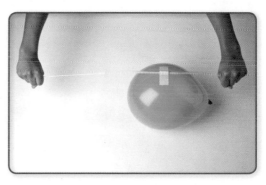

Thread a string through the straw. One partner holds each end of the string. Move the balloon to one end of the string. Let go of the balloon.

What happened to the balloon? How does the third law of motion explain what you observed?

Before You Move On

1. What are balanced and unbalanced forces?
2. Why is it easy to push a toy truck, but hard to push a real truck?
3. **Apply** How does a bed exert force on you while you are lying in it?

Conclusion

Motion can be described by an object's direction and speed, and also by its position. Motion is caused by forces. A force is a push or a pull. Forces can be contact or non-contact. Changes in motion are caused by unbalanced forces. How fast an object speeds up, slows down, or changes direction depends on the object's mass and the force acting on it.

Big Idea Newton's laws of motion describe how unbalanced forces on an object cause changes in the object's motion.

Vocabulary Review

Match the following terms with the correct definition.

A. **force**
B. **gravity**
C. **motion**

1. A push or a pull
2. A force that pulls things to the center of Earth
3. A change in position

Big Idea Review

1. **Explain** How are balanced and unbalanced forces related to each of Newton's laws of motion?

2. **Infer** A child and an adult are ice skating. The child pushes away from the adult. The child moves backward. What will happen to the adult?

3. **Predict** A feather falls off a bird in flight. The force of gravity acts on the feather. How does this force affect the feather's motion?

4. **Infer** Use Newton's first law of motion to explain why it is safer to wear a seat belt in a moving car.

5. **Develop a Logical Argument** Use Newton's second law of motion to explain why kicking a bag of soccer balls across a field takes more force than kicking one soccer ball across a field.

6. **Analyze** Two forces are acting on an object at rest. How can you tell if the forces are balanced or unbalanced?

Write About the Laws of Motion

Apply Concepts Think about a sport or an activity that you do at home. Describe how each law of motion is observed in the sport or activity.

CHAPTER 3 PHYSICAL SCIENCE EXPERT: KINESIOLOGIST

What Does a Kinesiologist Do?

William Sands is a kinesiologist. He uses his knowledge of forces, motion, and the human body to help top athletes and others perform at a higher level.

NG Science: What do you study?

William Sands: Most of my research time is spent on studying elite athletes, specifically gymnasts. I study just about all aspects of high performance in order to make better athletes. I analyze where the athlete makes mistakes in their routine. This helps me make the athlete's performance better in the future.

NG Science: What type of research have you done?

William Sands: I research the human body and the way it moves. For example, in the picture to the left, I am putting electrodes on a national champion archer. These electrodes are used to detect tiny electrical signals that are sent by his brain to the muscles in his arms which produces force. I study these tiny signals to determine just how elite athletes use their muscles to accomplish their amazing skills.

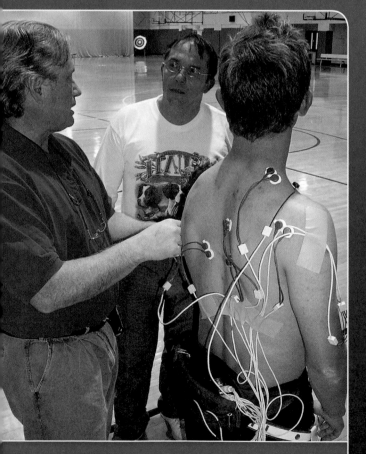

William Sands connects the archer to a computer to see if the archer is moving the muscles in his body correctly.

NG Science: What is your favorite thing about your area of research?

William Sands: The thrilling part of my research is when I see the athletes that I work with achieve their goals. I use science to help them, but the science is small compared to their dedication, determination, and commitment to their sport.

NG Science: What have you enjoyed most during the course of your career?

William Sands: I was lucky enough to coach several Olympians in women's gymnastics. I have now turned to science to help even more people with movement problems.

William Sands helped this gymnast train for her performance successfully.

NG Science: What type of training do you need to become a kinesiologist?

William Sands: To become a kinesiologist, you will need to study many kinds of science, as well as mathematics. You will also need to choose the type of body motion you find most interesting.

NG Science: Why is kinesiology important?

William Sands: If it moves, kinesiologists study it. Kinesiologists study motion so that everyone can move well; whether you want to dunk a basketball or simply rise from a chair.

121

BECOME AN EXPERT

Gymnastics: Forces in Action

Athletes in every sport learn how to exert the right **forces** to do well. In sports such as gymnastics, athletes use forces to control their own bodies.

Gymnasts use forces to perform different skills. Their hands and feet push off from the ground when doing tumbling. Their arms pull on the uneven bars and the parallel bars. As they exert these forces, each of the gymnasts are in **motion**. The movement of their bodies follows Newton's laws of motion.

Like all sports, gymnastics involves exerting the right forces at the right times.

force
A **force** is a push or a pull.

motion
Motion is a change in position.

TECHTREK
myNGconnect.com

Student
eEdition

Digital
Library

Some gymnastics skills require a gymnast to stay in one position. Men on the still rings perform strength moves. To keep his body in the correct position, a gymnast has to make sure that the forces on his body are balanced.

Gravity pulls on the gymnast while he is on the still rings. What happens if the gravity and other forces on the gymnast become too strong? If forces are unbalanced, and the gymnast pushes upward too hard or not hard enough, his body will not move in the correct way.

To hold this position, this gymnast needs to keep the forces on his body equal.

gravity
Earth's **gravity** is a force that pulls things to the center of Earth.

Speed and Direction

Gymnastics is not only about balancing and staying still. Gymnasts must sometimes move at high speeds. Speed is how fast an object moves. Gymnasts have to build speed when using the vault or tumbling. Gymnastics is also about moving.

The average speed of any object is calculated by dividing the distance the object goes by the time needed to go that distance. For example, a female gymnast runs about 27 meters (89 feet) before reaching the vault. She may run this distance in about 3 seconds. Her speed running toward the vault can be found using the following equation:

$$\text{speed} = \frac{27 \text{ meters}}{3 \text{ seconds}} = 9 \text{ meters/second}$$

A gymnast needs to be traveling at a high speed before hitting the springboard of the vault.

Gymnasts also have to change their speed. One way gymnasts change their speed is by moving faster. When a gymnast is performing on a pommel horse, he has to swing his legs around very quickly. But it is not possible for him to move his legs very fast at first. Instead, he starts moving them slowly. Then he speeds up.

On other apparatus, gymnasts also have to move more slowly. Whenever a gymnast dismounts an apparatus, he or she must come to a complete stop to "stick" the landing. But the gymnast is almost always moving before the dismount. The gymnast must slow down in order to stop moving.

Gymnasts also change direction. Obviously, gymnasts change directions whenever they run in a different direction. But gymnasts also change directions whenever they swing around in a circle on the high bar. Things moving in circles are always changing direction.

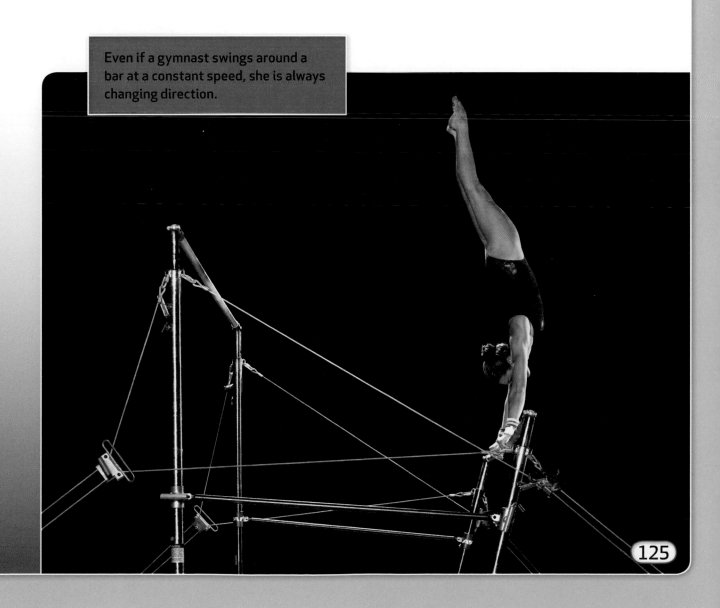

Even if a gymnast swings around a bar at a constant speed, she is always changing direction.

The First Law of Motion in Gymnastics

Before a gymnast starts moving, he or she is at rest. To start moving, the gymnast has to exert an unbalanced force. According to Newton's first law of motion, an unbalanced force will cause an object at rest to move. Often, that unbalanced force is the gymnast's feet pushing off the ground. But sometimes that force is a pull. Many times, men have to pull themselves up on a different apparatus to start their routines.

This gymnast had to exert unbalanced forces to move his body into the handstand position. But once he got into the position, he had to keep the forces balanced so that he would not fall.

An object in motion will stay in motion unless acted on by a greater force. You can see this part of Newton's first law of motion whenever a gymnast dismounts from the uneven bars. Right before a gymnast dismounts, she pushes off from one of the bars. Then she starts flying through the air. While she is in the air, gravity is pulling her down. She stops moving when she lands on the ground. The ground's force is equal to the force exerted by the gymnast.

When a gymnast is ready to dismount, she must push off from one of the bars.

After the gymnast pushes away from the bars, Earth's gravity is pulling her to the ground.

The Second Law of Motion in Gymnastics

When a gymnast wants to perform a small hop on the beam, she exerts only a small force. But when the gymnast wants to dismount off the beam, she exerts a large force.

She jumps high in the air. She is high enough to do twists or somersaults in the air before she lands.

How does the second law of motion help to explain this? To perform different moves, gymnasts have to exert different amounts of force.

TECHTREK
myNGconnect.com

Digital Library

A gymnast exerts forces of different sizes to perform different skills on the balance beam.

Remember, gymnasts create force by using their muscles. You have to be strong to be a gymnast. It doesn't matter how big or small you are; a lot of force must be exerted to perform these very difficult movements.

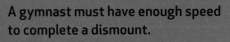

A gymnast must have enough speed to complete a dismount.

A gymnast exerts force to balance on the beam.

The Third Law of Motion in Gymnastics

Let's look at Newton's third law of motion in action. The third law states that for every action there is an equal and opposite reaction. Look at the photo below. The gymnast exerts forces on the equipment. At the same time, the equipment pushes back on the gymnast. These forces push him up into the air. Without these forces, gymnasts couldn't perform.

The pommel horse pushes on the gymnast with an equal and opposite force that keeps the gymnast's body in the air.

These forces are equal in size. They are also opposite in direction. Look at the photo below. A gymnast pushes on the vault. The vault pushes back with a force of equal size on the gymnast. This force sends the gymnast flying up off the vault.

Gymnasts train their bodies to move in precise ways.

CHAPTER 3
SHARE AND COMPARE

Turn and Talk How do balanced and unbalanced forces affect the way gymnasts perform their routines? Form a complete answer to this question together with a partner.

Read Select two pages in this section. Practice reading the pages. Then read them aloud to a partner. Talk about why the pages are interesting.

my SCIENCE notebook **Write** Write a conclusion that summarizes what you have learned about the different ways gymnasts use force. In your conclusion, restate what you think is the Big Idea of this section. Share what you wrote with a classmate. Compare what each of you wrote.

my SCIENCE notebook **Draw** Form groups of three. Have each person draw a picture that shows one of Newton's three laws of motion. Write labels that explain your drawing. Compare your drawing with your partners' drawings.

CHAPTER
4

WHAT ARE SIMPLE MACHINES?

Cutting a sandwich, riding a bike, opening a jar, walking up a ramp—what do all these have in common? They all involve simple machines that make work easier.

TECHTREK
myNGconnect.com

Student eEdition · Vocabulary Games · Digital Library · Enrichment Activities

Simple machines, such as the pulleys moving the cable car over the whirlpool in Niagara Gorge, Canada, make work easier.

134

135

After reading Chapter 4, you will be able to:

- Explain what work is and how work can be applied to objects. **WORK**
- Explain how simple machines change the amount of force or the direction of force needed to do work. **SIMPLE MACHINES**
- Describe how friction affects the amount of force needed to do work. **SIMPLE MACHINES**
- Compare the measures of different forces needed to move an object. **MEASURING FORCE**
- Describe how a spring scale works. **MEASURING FORCE**
- Identify and describe simple machines in common tools and household items. **EVERYDAY TOOLS**
- Science in a Snap! Identify and describe simple machines in common tools and household items. **EVERYDAY TOOLS**

CHAPTER 4

WHAT ARE SIMPLE MAC

Cutting a sandwich, riding a bike, opening a jar, walking up a ramp—what do all these have in common? They all involve simple machines that make work easier.

Simple machines, such as the pulleys moving the cable car over the whirlpool in Niagara Gorge, Canada, make work easier.

MACHINES?

SCIENCE VOCABULARY

lever (LE-vur)

A **lever** is a bar that turns against an unmoving point. (p. 141)

A hockey player uses a hockey stick as a lever.

inclined plane (IN-clined PLĀN)

An **inclined plane** is a flat surface with one end higher than the other. (p. 142)

This woman uses an inclined plane to make it easier to push the box into the truck.

screw (SKRŪ)

A **screw** is a bar that has an inclined plane wrapped around it. (p. 144)

This person is using a screw to hold pieces of wood together.

my
Science Vocabulary

inclined plane
(IN-clīned PLĀN)

lever
(LE-vur)

pulley
(PUL-lē)

screw
(SKRŪ)

wedge
(WEJ)

wheel and axle
(WĒL AND ACK-sel)

TECHTREK
myNGconnect.com

Vocabulary Games

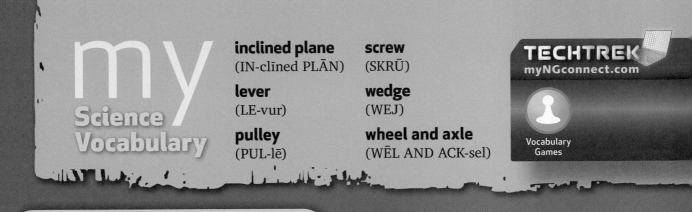

wedge (WEJ)

A **wedge** is a double inclined plane that can split objects apart. (p. 146)

The blade of a knife is a type of wedge.

wheel and axle (WĒL AND ACK-sel)

A **wheel and axle** is a large outer wheel attached to a smaller wheel called an axle. (p. 148)

A Ferris wheel has a wheel and axle.

pulley (PUL-lē)

A **pulley** is a grooved wheel with a cable or rope running through the groove. (p. 150)

The men use pulleys to raise and lower themselves on the outside of a building.

Work

Take a look at the soccer player. Is she doing work? You might be surprised by the answer. Work is using force to move an object over a distance. She is definitely doing work!

When the woman kicks the ball, she applies force to the ball. The ball then moves a certain distance. The distance the ball travels is the measure from where the girl kicked it to where it stops rolling.

Since the soccer player applied force to the ball by kicking it, and the ball moved because of the force, the soccer player did work by kicking the ball.

IS IT WORK?	YES OR NO
a pencil falling off a desk	yes
a pencil sitting on a desk	no
pushing a shopping cart across a room	yes
pushing against a brick wall that doesn't move	no

Before You Move On

1. What is work?
2. How can playing soccer be work?
3. **Apply** Think of any activity that you do at school. It might be opening a door or playing a drum. Explain whether that activity is work, and why.

Simple Machines

Simple machines exist in everyday objects around the home and in larger tools you might find at a construction site. You will also find them in sports, such as hockey. Simple machines are devices that make work easier.

Some simple machines make work easier by applying the force over a greater distance. This spreads out the work you need to do to move an object so that you don't have to do all of the work at one time. Other simple machines make a job easier by changing the direction of a force.

There are six simple machines: lever, inclined plane, screw, wedge, wheel and axle, and pulley. Each of these simple machines makes work easier in its own unique way.

A hockey stick is a type of lever.

Lever What simple machine could help you move something heavy? A lever might be able to take care of the job. Look at the seesaw below. A seesaw is a type of lever. The bar or board that you sit on has a fulcrum in the middle. The fulcrum is an unmoving part that changes the direction of the force. When you go down, your friend on the other end of the bar goes up. The lever changes the direction of the force so that it is easier to lift your friend up off the ground.

How It Helps When one end of the bar goes down or is pushed down, the other end goes up. The direction of the force is changed. In other words, by pushing down on one end of the bar, something is raised up on the other end.

bar

fulcrum

A seesaw makes the work of lifting someone easier.

Inclined Plane Lifting heavy objects from a lower place to a higher place can be very difficult. These people are trying to lift a heavy box into a truck. They could lift it straight up and put it in the truck, or they could use an **inclined plane** to make the work easier. An inclined plane is a flat surface with one end higher than the other.

How It Helps It changes how much force is needed to move something from a lower place to a higher place. An inclined plane allows work to be done using less force, though the distance traveled will be greater.

Lifting an object straight up takes a lot of force. If these people had an inclined plane, they would be able to use less force.

Inclined planes can be different lengths and heights. Some are steep and others are more level. The longer the inclined plane, the more gradual the increase in its height. The shorter the inclined plane, the sharper the increase in height.

It takes less force to push an object up a long inclined plane than it takes to push the same object up a short inclined plane. Which plane below would you rather use?

A long inclined plane makes getting the box into the truck a much easier job.

Using a short inclined plane requires more force to do the same amount of work than using a long inclined plane.

Screw You have probably seen screws in a toolbox, but have you ever seen one drill a huge hole at a construction site? Large screws called augers drill holes for construction projects. A screw is a bar that has an inclined plane wrapped around it. You can clearly see the inclined plane wrapped around this long bar. As the bar turns and goes downward, the inclined plane lifts soil out of the hole.

How It Helps To help understand how the inclined plane lifts the soil, imagine a spiral stairway in a lighthouse. The stairs go around, just like the inclined plane on this auger. You start at the bottom, and the stairway—the inclined plane—lifts you upward.

Screws come in many sizes. This enormous screw is used on a construction site.

Look at this jar. Two screws keep the lid on the jar—the ridges on the outside of the jar and the ridges inside the jar lid. The ridges are inclined planes that wrap around both areas. When the ridges are twisted around each other, the lid stays in place. You twist the lid to the right to tighten it.

You twist it to the left to loosen it. This is the same motion the auger makes when it drills a hole.

Most screws are used as fasteners. A wood screw might fasten two pieces of wood together. Bolts are another kind of screw. They are used to fasten pieces of metal together.

This screw holds things together. The ridges of the inclined plane create friction with the wood so that the screw can't easily be pulled out.

The jar's lid stays tight keeping the food inside the jar fresh.

Wedge What can you observe in the photo of the sailboat? Notice how the front of the sailboat seems to be cutting through the water? The front of the boat uses a simple machine called a **wedge** . Boats that are made to go swiftly through the water have a long thin shape. The hull, or main body, of the boat is a wedge. It pushes water to the side as it moves forward. A wedge is a double inclined plane with the ability to move as it does work.

What common kitchen tool uses a wedge to cut fruits and vegetables? The blade of a knife is a wedge. People use knives every day to cut fruits and vegetables. Most wedges, such as knives and shovels, are made to cut through surfaces.

TECHTREK
myNGconnect.com

Digital
Library

The hull of this sailboat is a wedge that allows the boat to move easily through the water.

wedge

How It Helps A knife cuts through food. Notice how the knife pushes aside the two parts of the food. The knife blade—the wedge—changes the downward motion of the force to a sideways motion that pushes the food apart.

The knife's long, thin wedge easily cuts through food.

Wheel and Axle   This Ferris wheel is another kind of simple machine called a **wheel and axle**. It has cars attached to the wheel that people can ride in. A wheel and axle is a large outer wheel attached to a smaller wheel called an axle. Examples include doorknobs and wheels and axles on cars, buses, and trains.

How It Helps   In a wheel and axle, one part cannot turn without the other part turning. Applying force to the large wheel results in a more powerful force at the axle.

When the axle turns, the outer wheel of this Ferris wheel—and all of its cars—turn too.

Friction What do you observe in the photos of the wagons? Based on your observations, which wagon would you rather be pulling? Friction affects the amount of work it takes simple machines to do their job.

If a wagon is pulled over a rough surface, it takes more effort than if it is pulled over a smooth surface. The rough surface causes more friction than the smooth surface does. So, more force is needed to pull the wagon on the rough surface than on the smooth surface.

FRICTION WITH SMOOTH AND ROUGH SURFACES

Here the ground is smooth. There is less friction between the wheels and the ground.

The leaves on the ground make the surface rough. There is more friction between the wheels and the ground.

Pulley These window washers use pulleys to move up and down the building. A pulley is a grooved wheel with a cable or rope running through the groove.

When the window washers want to move up the building, they pull on the rope. Since the rope is attached to a pulley, it makes the work of moving up the building easier. This is because the pulley changes the direction of the force. The window washers pull down on the rope, and they go up the building.

Pulleys control the height of each window washer on an office building.

Usually, two or more pulleys are used together. When pulleys are used together, it takes less force to lift a load.

How It Helps A pulley is a machine that is used to lift objects. Lifting an object directly takes more force than it takes to lift the same object with a pulley. By pulling down on the rope, you cause an object at the other end of the rope to rise.

A pulley has two parts: a grooved wheel and a rope.

Before You Move On

1. Identify a common example of a lever.
2. How are an inclined plane and a screw alike? How are they different?
3. **Generalize** How do simple machines make work easier?

Measuring Force

You can measure the amount of force needed to do work. The amount of force needed to move an object is measured in newtons. Look at the picture below. A spring scale is being used to hold up a bunch of grapes. The spring scale uses about 1 newton of force to hold up the grapes.

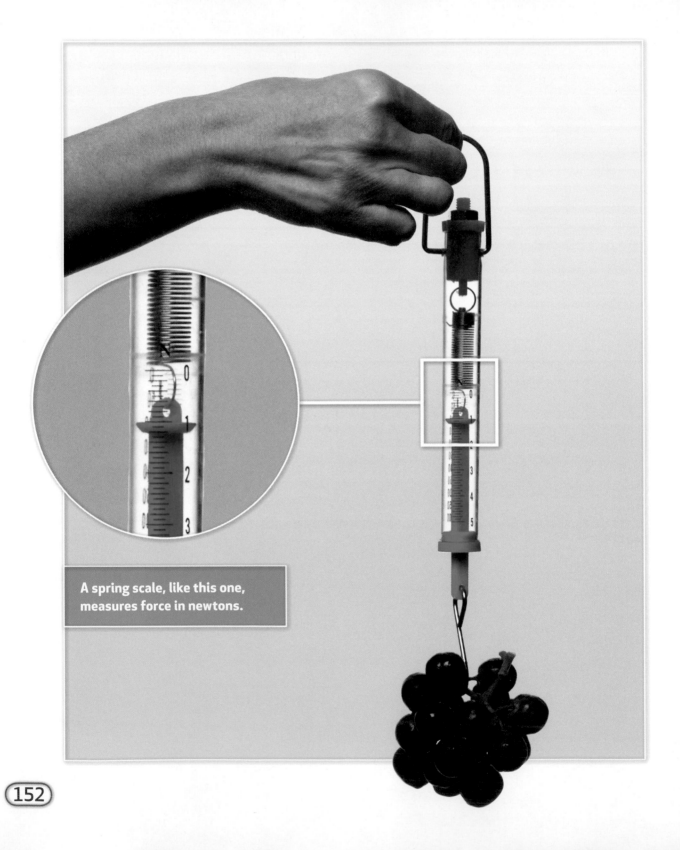

A spring scale, like this one, measures force in newtons.

Simple machines need to use less force to do the same amount of work. Look at the examples below. Did it take more newtons to lift the bottle in the top picture or pull the bottle in the bottom picture?

Without an inclined plane, it took more newtons to lift the bottle. It takes much less newtons of force to pull the bottle when the inclined plane is used. The inclined plane made the work easier, so less force was needed to move the bottle.

It takes less force to move the bottle using a simple machine.

Before You Move On

1. What unit of measure would you use to tell how much force is needed to lift something?
2. What effect would using a simple machine have on the amount of force it takes to move an object?
3. **Infer** Why might it be useful to know how much force it takes to lift an object?

153

Everyday Tools

You use simple machines in your life every day. Did you know that every time you use a doorknob to open a door you are using a simple machine? A doorknob is a wheel and axle.

You apply force to the large, outer wheel. That causes the axle to turn. When you turn the doorknob, the axle inside turns with enough force to open the door.

The axle of the doorknob is the part that goes through the door from one side to the other.

A wheelbarrow makes work easier. It is made of a few simple machines. A wheelbarrow has a wheel and axle. It also has a lever. The end of the bar where the wheel is located acts as the fulcrum.

Lifting something heavy in a wheelbarrow takes less force than it would take to lift it in your arms.

fulcrum

Unlike a seesaw, whose fulcrum is in the middle of the bar, a wheelbarrow's fulcrum is at the end of the bar.

Have you ever seen someone cut wood with an ax? An ax has a wedge and a lever. The wedge on the ax cuts through the wood. The person uses the handle as a lever to force the wedge through the wood.

A shovel is also made of simple machines. Its handle is a lever. Its blade is a wedge. The lever and the wedge work together. They make digging a hole easier.

This shovel makes the work of moving dirt easier.

This man uses an ax to split wood.

A scissors uses simple machines to cut paper.

Enrichment Activities

wedges

Scissors are also made of simple machines. The blades of the scissors are wedges. The handles of the scissors are levers. You use the levers to move the wedges together and apart. When the wedges come together, they can cut through paper and other materials.

Science in a Snap! Keep It Simple

Find an everyday object.

Analyze the object.

What simple machines are in the object?

Before You Move On

1. What type of simple machine is a doorknob?
2. Describe how scissors work.
3. **Apply** What simple machines make up a stapler?

157

SIMPLE MACHINES
AND AIRPLANES

Many inventors and engineers experimented with aircraft before the Wright brothers' famous flight in 1903. But the Wright brothers' first airplane, called the Wright flyer, was unique:

1. It was powered by an engine.
2. Its direction could be controlled.
3. It had a sustained flight.

These three things had never been done at the same time before. The Wright brothers were familiar with simple machines because they ran a bicycle shop and designed bicycles. Simple machines were a key part of the first airplane.

This hand-colored photograph was taken in Kitty Hawk, North Carolina, right after the Wright flyer lifted off the ground on December 17, 1903.

Wheel and Axle The first plane had wheels and axles made out of bicycle hubs. This simple machine allowed the plane to run down a rail and launch into the air.

Pulley The engine of the plane used a pulley. The pulley was linked to the propellers and the motor.

Lever Another simple machine, the lever, was used in two places on the airplane. The pilot accelerated the plane by pulling a lever. A separate lever controlled the front of the plane.

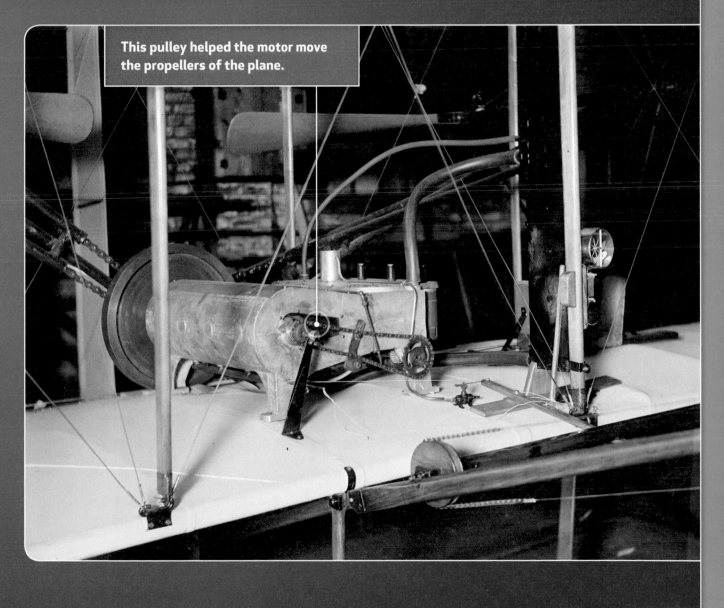

This pulley helped the motor move the propellers of the plane.

Conclusion

Simple machines make work easier by changing the amount of force needed or by changing the direction of force. All six of the simple machines—lever, inclined plane, screw, wedge, wheel and axle, and pulley—can be found in everyday objects.

Big Idea Simple machines change the amount of force needed to do work.

Vocabulary Review

Match the following terms with the correct definition.

A. **inclined plane**

B. **wedge**

C. **wheel and axle**

D. **lever**

E. **pulley**

F. **screw**

1. A bar that turns against an unmoving point
2. A flat surface with one end higher than the other
3. A bar that has an inclined plane wrapped around it
4. A double inclined plane that can split objects apart
5. A large outer wheel attached to a smaller wheel
6. A grooved wheel with a cable or rope running through the groove

Big Idea Review

1. **Explain** How is playing the violin an example of doing work?

2. **Compare and Contrast** How are the wedge and inclined plane alike and different?

3. **Apply** When have you used a lever?

4. **Explain** Suppose you were attempting to split a piece of paper in half. Why would scissors be a better option to choose than a knife?

5. **Predict** How would the force needed to move a wagon change if the wheels were removed?

6. **Evaluate** If you were stuck on a desert island, what two simple machines would you want with you? Tell why.

Write About Simple Machines

Explain What is happening in this photo? What simple machine would make this work easier? Give details in your answer.

PHYSICAL SCIENCE EXPERT: INDUSTRIAL DESIGNER

Bicycles use a number of simple machines. Bicycle designer Roxy Lo uses her understanding of simple machines on a daily basis to design bicycles that use force efficiently.

NG Science: What is an industrial designer, and what do you do?

Roxy Lo: Industrial designers are a special kind of engineer. First and foremost, they must be experts on the materials they use to design their products. With this knowledge, they must then combine the science of the materials with their artistic vision of what they want the product to look like. What I do specifically is design unique bicycle frames out of a lightweight material called carbon fiber, while making sure it will not negatively impact the inner workings of the bicycle.

A bicycle has many moving and nonmoving parts. Many of these are simple machines that must work together seamlessly to allow the bicycle to move properly. When designing and building my frames, I always must make sure my design does not interfere with how those simple machines work together.

TECHTREK
myNGconnect.com

Digital Library

Roxy Lo is an industrial designer who designs bicycle frames.

Pulley

Wheel and Axle

NG Science: What's a typical workday?

Roxy Lo: I have one office at work and one at home, so I can work whenever I feel creative. Sometimes I draw designs for ten hours a day. I love my job, and each time I come up with a new design, I am reminded that I am doing something fun!

NG Science: How did you get into industrial design?

Roxy Lo: I have always enjoyed making things. I have designed all kinds of products. However, I had always thought that it would be fun to design something mechanical. Then one day, I had a chance to develop a bicycle using carbon fiber in a special process. I tried it and I knew that I wanted to design more bikes.

NG Science: What did you have to learn about to be able to do your job?

Roxy Lo: In college, I took all kinds of classes, from technical drawing to manufacturing processes. I had to take a lot of math and science courses to get a good understanding of how things work. I also had to learn about proper communication. An industrial designer needs to be able to communicate her vision to engineers or other industrial designers, from the first idea to the final project.

NG Science: What is the most important part of your job?

Roxy Lo: Understanding how to make innovative designs manufacturable and easy to use is very important. I've traveled all over the world to learn about how things are made. People make products differently all over the world, and it is always interesting to learn new ways to design new products.

NG Science: What is your favorite thing about your job?

Roxy Lo: My favorite part of my job is seeing other people enjoy the products I design. If you like how something looks and works, you use it more.

BECOME AN EXPERT

Become an Expert: Bicycles

You probably know that bicycles can be a lot of fun. However, what you may not know is that bicycles are not a new form of transportation.

Early bicycles had wheels made of metal, which gave the bicyclist a bumpy ride.

The earliest bicycle was invented by Baron Karl von Drais de Sauerbrun in Germany in 1817. On one of the first bicycle rides ever, he rode it for 14.5 kilometers (9 miles). It was made of wood and he walked it along with his feet against the ground. Changes in design quickly followed. Soon, the frames and wheels were made from iron. Because of how bumpy the ride was, these bicycles were nicknamed "boneshakers"! Over the next hundred years, rubber wheels and wire spokes for tires helped improve the bicycles by giving them a smoother ride.

TECHTREK
myNGconnect.com

e
Student
eEdition

Digital
Library

When you look closely at bicycles, you can see that bicycles involve a combination of simple machines , such as **wheels and axles** and **pulleys** .

Bicycles move quickly and cover a great distance with only foot power. Look at the picture below. When you hop on a bicycle, your feet exert force on the pedals. This provides force to the gear wheel. The gear wheel applies force to the chain, and the chain, in turn, applies force to the back wheel, which moves you forward on your bicycle.

TECHTREK
myNGconnect.com

Many simple machines work together to make the bike move more easily.

Digital
Library

back wheel
(wheel and axle)

pedal
(wheel and axle)

gear wheel and chain
(pulley)

wheel and axle

A **wheel and axle** is a large outer wheel attached to a smaller wheel called an axle.

pulley

A **pulley** is a grooved wheel with a cable or rope running through the groove.

Inclined Plane Have you ever ridden a bicycle up a hill? Each hill is an inclined plane. The slope of each **inclined plane** is different. Some are long and gradual. Some are short and steep.

Bicyclists use inclined planes to get to higher and lower places.

This bicyclist is going down a hill which is a type of inclined plane.

Bicyclists are using this inclined plane to get to the top of the mountain in this course.

inclined plane

An **inclined plane** is a flat surface with one end higher than the other.

Wheel and Axle

The pedals on a bicycle use a wheel and axle, made of a large wheel and a smaller wheel that acts as an axle. When you push the pedals, the smaller wheel turns. Since it is a small wheel, you need less force to move it. When the smaller wheel turns, it makes the larger wheel turn too. And when the wheels turn, the bike moves, and off you go!

The wheels and pedals of a bicycle are simple machines.

Screw Screws fasten parts of bicycles together. Screws fasten the axles to the wheels. Screws can also attach equipment, such as a water bottle holder, to a bicycle.

Unlike some screws used in construction, bicycle screws are made to fit each bicycle part. When fully screwed in, the head of a bicycle screw lies inside the frame so that it does not create any friction in the wind. Less friction from the wind means a faster, smoother bicycle ride for you.

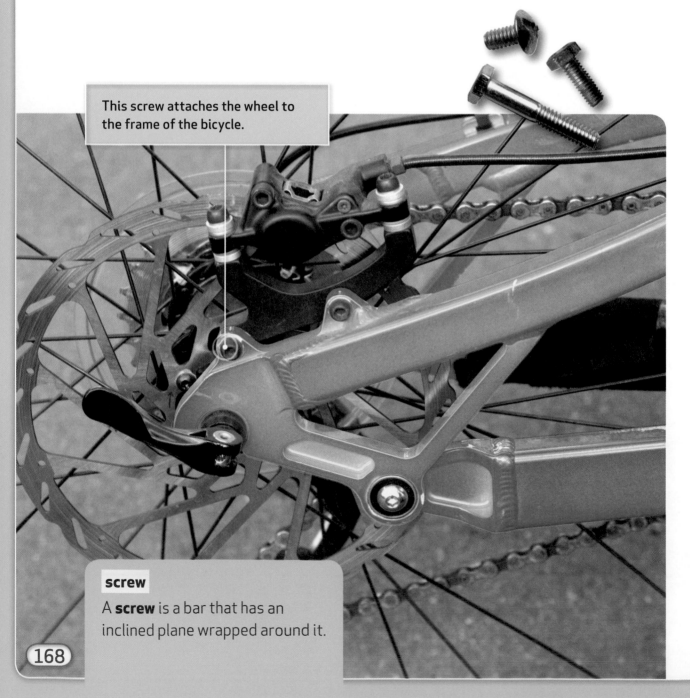

This screw attaches the wheel to the frame of the bicycle.

screw

A **screw** is a bar that has an inclined plane wrapped around it.

Pulley Bicycles use a special pulley. It consists of the bicycle chain, and a gear wheel. The force is applied to the front end of the chain when the bicyclist presses down on the pedal. Pulleys normally lift objects upward. In this case, the pulley "lifts" the bicyclist, but not upward. Instead, the pulley moves the rider forward whenever he or she applies force to the pedals.

To reduce the amount of friction on the bicycle chain, bicyclists add a special liquid to the chain. This makes the chain move more smoothly, and makes for a better ride.

This gear wheel works a pulley. The chain takes the place of the rope that would be part of a normal pulley system.

BECOME AN EXPERT

Lever The brakes and gear shifts on a bicycle are types of **levers**. By pulling on the lever to brake or change gears, the other end of the lever moves in the opposite direction. Levers change the direction of a force.

Bicycles are engineered to be aerodynamic. This means they move through the air with as little friction as possible. One technique that bicycle designers use is to make the brakes smaller on the bicycle. The less clutter on the bicycle frame, the faster the bicycle can go.

The rider applies force to the brake lever to stop the bicycle.

Notice the smooth, curved surfaces of the brakes. The design prevents them from creating too much friction.

lever
A **lever** is a bar that turns against an unmoving point.

Wedge In some races, bicyclists participate as part of a team. Team members lend each other parts and help set the pace for one another. They can also act as a human **wedge** for one another. With an aerodynamic helmet and bicycle, and by crouching down low, bicyclists can slice through the air as they pedal. This is similar to the way the hull of a boat acts as a wedge in the water. The bicyclist who immediately follows can ride in the draft of the first bicyclist , and use less energy.

Bicycles are made up of many simple machines that makes riding them easier and more fun!

The smooth surface and curves of this helmet help the bicyclist cut through the air.

wedge
A **wedge** is a double inclined plane that can split objects apart.

CHAPTER 4
SHARE AND COMPARE

Turn and Talk How are simple machines used in a bicycle? Form a complete answer to this question together with a partner.

Read Select two pages in this section. Practice reading the pages. Then read them aloud to a partner. Talk about why the pages are interesting.

my SCIENCE notebook

Write Write a conclusion that summarizes what you have learned about bicycles and simple machines. State what you think is the Big Idea of this section. Share what you wrote with a classmate. Compare your conclusions. Did you recall which simple machines are used in bicycles?

my SCIENCE notebook

Draw Draw a picture of something in your house that is made up of simple machines. Add labels to name the simple machines. Share your drawing with a classmate. Tell how you use the machine and how it makes work easier. Compare how your drawings are alike and different.

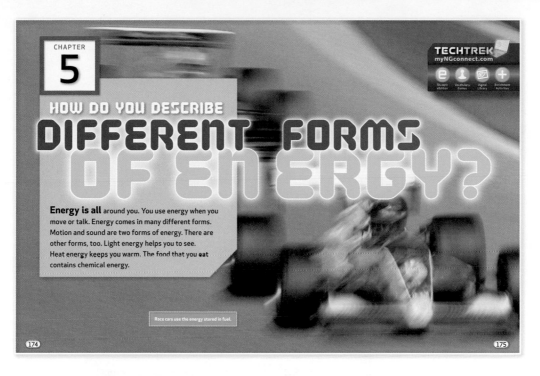

HOW DO YOU DESCRIBE
DIFFERENT FORMS
OF EN ERGY?

Energy is all around you. You use energy when you move or talk. Energy comes in many different forms. Motion and sound are two forms of energy. There are other forms, too. Light energy helps you to see. Heat energy keeps you warm. The food that you eat contains chemical energy.

Race cars use the energy stored in fuel.

After reading Chapter 5, you will be able to:

- Identify and describe different forms of energy, including mechanical, sound, light, heat, and chemical. **MECHANICAL ENERGY, SOUND, LIGHT, HEAT, CHEMICAL ENERGY**

- Explain that energy is the ability to cause motion or to create change. **ENERGY AND WORK, MECHANICAL ENERGY, CHEMICAL ENERGY**

- Describe the transmission, reflection, and absorption of sound. **SOUND**

- Describe the transmission, reflection, and refraction of light. **LIGHT**

- Explain that heat moves from warmer objects to cooler ones. **HEAT**

- Science in a Snap! Describe the transmission, reflection, and refraction of light. **LIGHT**

HOW DO YOU DESCRIBE DIFFERENT OF En

Energy is all around you. You use energy when you move or talk. Energy comes in many different forms. Motion and sound are two forms of energy. There are other forms, too. Light energy helps you to see. Heat energy keeps you warm. The food that you eat contains chemical energy.

Race cars use the energy stored in fuel.

TECHTREK
myNGconnect.com

Student
eEdition

Vocabulary
Games

Digital
Library

Enrichment
Activities

FORMS ERGY?

SCIENCE VOCABULARY

potential energy
(pō-TEN-shul EN-er-jē)

Potential energy is stored energy. (p. 180)

> The blue car has more potential energy because of its position at the top of the ramp.

kinetic energy
(ki-NET-ik EN-er-jē)

Kinetic energy is the energy of motion. (p. 180)

> The cars are in motion. Their potential energy turned into kinetic energy.

my Science Vocabulary

chemical energy
(KEM-i-kul EN-er-jē)

mechanical energy
(mi-CAN-i-kul EN-er-jē)

kinetic energy
(ki-NET-ik EN-er-jē)

potential energy
(pō-TEN-shul EN-er-jē)

TECHTREK
myNGconnect.com

Vocabulary
Games

mechanical energy
(mi-CAN-i-kul EN-er-jē)

The **mechanical energy** of an object is its potential energy plus its kinetic energy. (p. 180)

This toy's mechanical energy is the sum of its potential energy plus its kinetic energy.

chemical energy
(KEM-i-kul EN-er-jē)

Chemical energy is energy that is stored in substances. (p. 194)

Your body runs on the chemical energy stored in food.

Energy and Work

Energy and work might make you think of raking leaves, sweeping the floor, or taking out the garbage. In science, work and energy have special meanings. Work is when you apply a force to move an object. Energy is the ability to do work or cause a change.

You can see that this girl is doing work. She pushes and pulls on the rake to gather the leaves into a pile. She applies a force to move objects (the rake and leaves). How is energy involved? The girl is able to do this work because she has energy. She has the ability to cause a change to the scattered leaves.

These leaves were scattered over the grass, but now most of them are in a pile. How is the girl using energy to do work?

Every time you move, you use energy. Sometimes, you transfer some energy to something else. That's what this kicker is doing to this ball—he's transferring energy to the ball. That energy puts the ball in motion. The energy changes the ball's position. The ball moves. When it lands, it's going to do work. How? It's going to bend some blades of grass. Anything can have energy and do work.

When you kick a football, you give energy to it to make it move. You apply work to it.

Before You Move On

1. What is energy?
2. How are work and force related?
3. **Apply** You kick a soccer ball to a friend. Explain how you have transferred your energy.

Mechanical Energy

The cars on this ramp have stored energy, which is also called **potential energy**. The cars have the ability to do work, but they aren't doing it at the moment. The blue car is higher off the ground than the red car. The blue car has more potential energy because it has a greater distance to travel before it reaches the top of the table.

When you let go of the cars, their potential energy turns into **kinetic energy**, or the energy of motion. All objects that are in motion have kinetic energy.

The **mechanical energy** of the cars is their potential energy plus their kinetic energy.

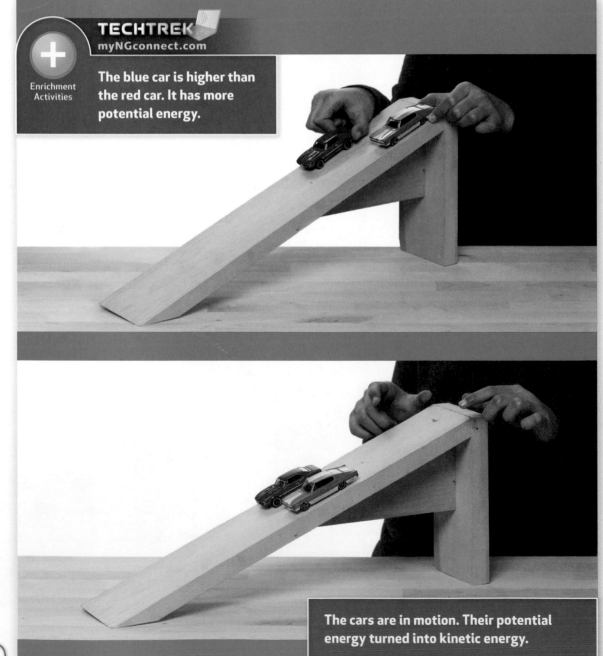

TECHTREK
myNGconnect.com

Enrichment Activities

The blue car is higher than the red car. It has more potential energy.

The cars are in motion. Their potential energy turned into kinetic energy.

This toy also has potential energy. There is a spring inside. When you turn the knob, you tighten the spring. The tighter it gets, the more potential energy the toy has.

What happens when you release the knob? The spring starts to loosen.

The potential energy turns into kinetic energy and the toy moves. Now it has energy of motion. The toy has some potential energy until the spring has fully loosened. The potential energy plus the kinetic energy is the mechanical energy of the toy.

Before You Move On

1. What is the mechanical energy of an object?
2. When does potential energy turn into kinetic energy?
3. **Infer** Look at the photos of the toy cars going down the ramp. When do you think the cars have the most kinetic energy?

Sound

Have you ever heard someone play the piano? The photo on this page shows the inside of a piano. A hammer strikes a set of wires, or strings. The wires vibrate. This vibration travels. Listeners then hear the sound of the piano.

Sound is another form of energy. It is energy you can hear. Sound travels in waves. Objects make sound waves when they vibrate. The waves travel away from the object. They spread out in all directions. They move through the air, liquids, and solids.

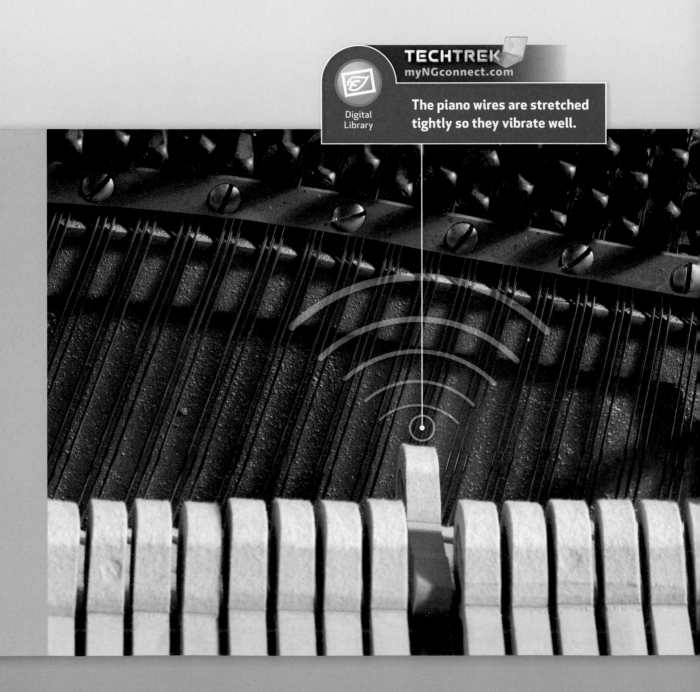

A piano makes many different sounds. These sounds can have different pitches. Pitch describes how high or low a sound is. A piano makes high-pitched sounds and low-pitched sounds. It can make these different sounds because each wire is a little different.

Short, thin wires produce the piano's high-pitched sounds. Because the wires are short and thin, they vibrate quickly, so we hear a sound with a high pitch.

Thick, long strings produce the piano's low-pitched sounds. Because the wires are thick and long, they vibrate slowly to produce a sound with a low pitch.

Short, thin piano wires vibrate quickly. They make sounds that have a high pitch.

Long, thick piano wires vibrate slowly. They make sounds that have a low pitch.

Reflection Some animals, such as bats and dolphins, use sound waves to find their way around and locate food. Look at the photo below. This bat is using sound waves to tell exactly where the moth is. The bat is able to do this because sound waves reflect, or bounce off of, surfaces.

Echoes are reflected sound waves. You hear an echo when the reflected waves bounce back to your ears. In this picture, the sound waves reflect off the moth and back to the bat. The bat is using echoes to find its next meal.

TECHTREK
myNGconnect.com

Digital
Library

The bat uses reflected sound
waves to find food.

Absorption Think of a library. It's carpeted. It's full of books, which are made of paper and cardboard. Those materials absorb sound energy well. Sound waves do not bounce off of them. Instead, sound waves go into the materials. The materials absorb some of the energy, so the vibrations become weaker. Finally, the sound waves fade. Libraries seem quiet because the materials in the library are absorbing the sound waves instead of reflecting them.

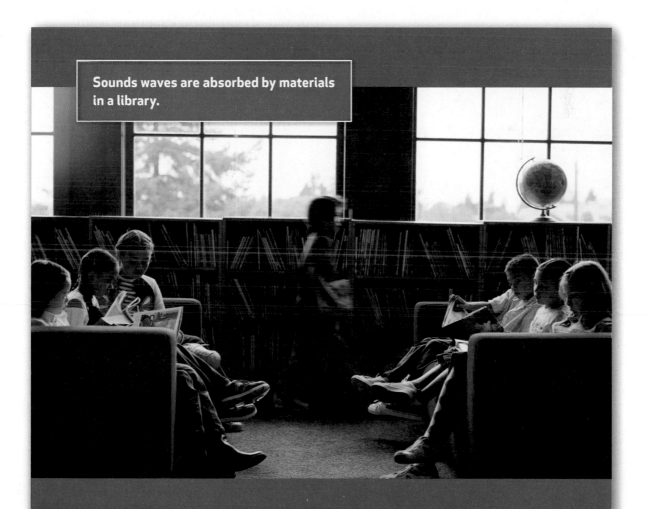

Sounds waves are absorbed by materials in a library.

Before You Move On

1. What causes a sound wave?
2. A thick piano wire and a thin piano wire are vibrating. Compare the sounds they make, and tell why they are different.
3. **Generalize** Tell why a gymnasium would not be a good place to have a quiet library.

Light

Have you ever noticed how one light bulb can light up an entire room? How does that happen? Light energy travels in waves. You can't see the waves, but you can see the light. Light travels away from its source in every direction from the light bulb to all parts of the room.

Light travels in a straight line. When you see an object, light is reflecting, or bouncing, off of that object. The light moves in a straight line to your eyes. You can only see the object when light reflects off of it into your eyes. When the straight path of the light is blocked, a shadow forms.

If something blocks the path of the light waves, such as your body or your hand, a shadow forms.

Look at the building in this photo. Based on your observations, what can you infer about the building's windows? The building's windows are like mirrors. They are reflecting the light.

When light waves bounce back from a mirror, they make a reflection that you can see. Because the building has so many mirrors on the outside, you can see a reflection of all the buildings across the street and of the sky.

Light waves from a nearby building reflect off of the mirrored building. They make a reflection.

Refraction What do you notice about the stem of this flower? It looks bent at the surface of the water. It looks bent because the light waves travel at one speed through the water, and at another speed through the air.

Light travels through clear materials. It bends, or refracts, when it passes from one material into another. It bends because the speed of light changes.

Light waves bend when they pass from the water to the air. The stem seems to have two parts because of the way the light waves bend.

Look at the prism on this page. When light passes through a prism, different wavelengths of light are bent by different amounts. This bending causes the different colors of light to separate. That is why you see a rainbow when the light exits the prism.

Pour water into a clear cup. Place a straw in the water.

Look at the straw through the side of the cup.

Describe what you see.

Before You Move On

1. What causes a shadow?
2. When does light refract?
3. **Infer** You can see a reflection of yourself in a shiny spoon. What can you infer about what the spoon is doing to the light waves?

Heat

Why does popcorn pop? It's all about heat. The popper heats the popcorn seeds. When the material inside the seeds gets hot, it bursts out of the hard outer shell. That's what makes the popping motion. But why does the inner material burst?

The particles that make up matter are always moving. Even the particles in solid materials, such as wood or ice, are moving. The particles have energy because of their motion. The energy of the moving particles is heat energy. The faster the particles of a substance move, the more heat energy they have.

When the particles inside the popcorn seeds are moving very fast, the shell can no longer contain the material. The "explosion" happens, and you get a snack.

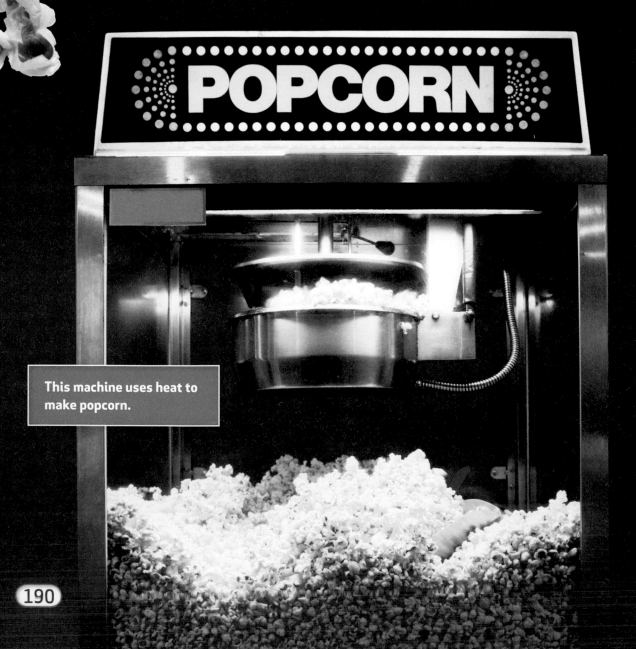

This machine uses heat to make popcorn.

The particles in the pool water are not moving as fast as the particles in the hot popcorn. In this photo, the temperature of the water is being measured. Temperature is the measure of how hot or cold an object is. When particles move slowly, the object or substance has a lower temperature. The swimmers probably hope that the particles in the water are not moving too slowly. They won't want to swim in cold water!

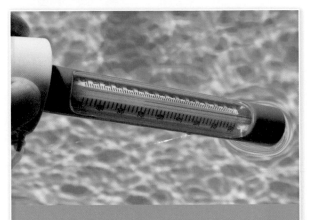

A person uses a thermometer to tell the temperature of the pool water.

Particles in cold substances, such as pool water, move slowly compared to particles in hot substances.

What do you think is happening in the photo? What happens when two objects that have different temperatures touch?

Heat energy flows between objects that have different temperatures. You can predict the direction that heat energy will flow. What happens when you pick up a warm mug? Your hands begin to feel warm. Your hands are cooler than the mug. Heat energy flows from the mug to your hand.

What happens when you pick up a snowball? Your hands begin to feel cold. They feel cold because heat is flowing away from your hands to the snowball. Heat always flows from warmer objects to cooler objects. The particles in the cooler object start to move faster. The object warms up. The warmer object cools down. Heat flows until the objects reach the same temperature.

Heat energy flows from your body to the cold air when it is cold outside. Heat energy also moves from a hot cup of cocoa to your hands.

Heat energy is transferred in different ways. If objects are touching, heat flows by conduction. Conduction happens when heat moves between two objects that touch. Heat energy moves by conduction when you place a pan on a hot stove. Look at the pan on this page. First, the heat energy moves from the stove's burner to the pan. Then, the heat energy moves from the pan to the water, and then to the broccoli.

CONDUCTION

Heat flows by conduction from the burner to the pan to the water to the broccoli.

Before You Move On

1. What is heat energy?
2. How does temperature relate to the speed of particles?
3. **Predict** In what direction will heat flow when cold water is added to a hot pan?

Chemical Energy

Your body runs on energy. Every time you take a bite of food, you are fueling your body. What kind of energy can your body use? The answer is **chemical energy**. Chemical energy is energy that is stored in substances. It is a type of potential energy because it is stored energy.

When you eat, the chemical energy from the food is stored in your body. Your body changes it into other types of energy. Some of the chemical energy becomes heat energy that keeps your body warm. Chemical energy changes into mechanical energy when you move. Your body also uses the energy in food to build the nutrients you need to grow.

Our bodies use the chemical energy stored in food.

Chemical energy can change into other types of energy. Sometimes burning causes this change. Natural gas is a fuel that contains chemical energy. What type of energy do you observe when a natural gas burner is lit? You see light energy. If you were standing nearby, you would feel heat energy. And, of course, the heat energy heats the pan sitting on the burner. Some of the chemical energy in the natural gas becomes light energy. Some becomes heat energy.

The chemical energy stored in natural gas is released when it burns.

Cars also need fuel to work. Most cars use a fuel called gasoline. Gasoline contains chemical energy. Burning gasoline changes the chemical energy into heat energy and light energy. Some of the heat energy can be used to do work.

In a car engine, heat energy changes into mechanical energy. It makes parts of the engine move. Then, the moving engine parts make the car move. Some of the heat energy is not used to do work. It makes the engine warm.

People fill their gas tanks with fuel at a gas station. The car engines burn the fuel.

Even if energy changes form, the total amount of energy stays the same. Think of the car engine. Chemical energy changes to heat energy. Some of the heat energy changes into mechanical energy. But no energy is lost.

TECHTREK
myNGconnect.com

Digital Library

Fuel is burned inside a car engine. Some of the chemical energy in the fuel changes into heat energy. Then, some of the heat energy changes into mechanical energy.

Before You Move On

1. Give an example of how energy can cause motion.
2. What happens to the chemical energy in gasoline when it is burned?
3. **Apply** Where does the heat energy and light energy of a candle come from?

NATIONAL GEOGRAPHIC

SEARCHING WITH SOUND IN EGYPT

The city of Alexandria, Egypt, is more than 2,000 years old. It has a rich history, but much of the ancient city was lost. Now people are finding clues to its past in an unexpected place—underwater.

Parts of the city were covered by sand and rising seas. Today, underwater archaeologists use sound energy to search the sea near the modern city. They are finding the remains of a great city.

Today, some of modern Alexandria's buildings stand nearly on the shore. Some of ancient Alexandria's buildings lie under the water.

A diver recovers a statue of an ibis, a bird that was sacred to one of the Egyptians' gods.

One of the tools underwater archaeologists use is sonar. Sonar uses reflected sound waves to find objects underwater. The archaeologists use sonar to map the bottom of the sea. They send out sound waves to the seafloor.

Then a computer makes a picture of what lies at the bottom using the echoes that return. In Alexandria, the archaeologists used the sonar pictures to find spots that looked like they might be parts of old buildings. Then, they sent divers to carefully uncover valuable pieces of the past.

Using sonar, scientists have identified where ancient Alexandria's buildings were. They are shown in yellow on this map.

A gold ring lies on the seafloor where ancient Alexandria once stood. The artist carved a bird into the beautiful blue-green stone.

Conclusion

Energy comes in many different forms. All forms of energy can do work or cause change. An object may have different forms of energy at the same time. It may transfer its energy to another object. Or, an object's energy may change to a different kind of energy.

Big Idea Energy comes in different forms and can be used to do work or cause change.

Mechanical Energy

Heat Energy

Chemical Energy

Light Energy

Sound Energy

Vocabulary Review

Match the following terms with the correct definition.

A. **potential energy**
B. **kinetic energy**
C. **mechanical energy**
D. **chemical energy**

1. An object's potential energy plus its kinetic energy
2. Energy that is stored in substances
3. Stored energy
4. The energy of motion

Big Idea Review

1. **Define** What is mechanical energy?

2. **Explain** On what does the kinetic energy of an object depend?

3. **Summarize** What happens when light travels through water?

4. **Predict** Frozen vegetables are added to a pot of boiling water. In what direction will heat energy flow?

5. **Draw Conclusions** Why does a car stop moving when it runs out of gasoline?

6. **Explain** How do bats locate food?

Write About Mechanical Energy

Describe Explain why a tennis ball held above your head has more potential energy than a tennis ball held at your waist.

PHYSICAL SCIENCE EXPERT: URBAN PLANNER

What Does an Urban Planner Do?

Do you know where the water that flows out of the faucet comes from? How does it get to your house or school? Who figures all of that out? Making sure people have plenty of water is just one part of an urban planner's job. As an urban planner, Thomas Culhane makes decisions about water and energy usage. Urban planners work to make towns and cities nice places to live. They plan where to put schools, parks, and roads. They also help find ways to use energy and other resources wisely and to control pollution.

Thomas Culhane shows a future explorer how to build his own solar hot water heater.

TECHTREK
myNGconnect.com

e
Student
eEdition

Digital
Library

What inspired Culhane? When he was 14 years old, Culhane saw the plans for a model community. It would use renewable resources for its energy sources. The community was never built, but Culhane never forgot the idea.

Urban planners must understand what people need and how they use resources, such as energy. Culhane has lived in the United States, Indonesia, Guatemala, and Egypt. Living in different places helps him see how people get food, water, and shelter. He started Solar CITIES to help people in poor parts of Cairo, Egypt. The people did not have warm water in their homes. Culhane's group helped people build water heaters that used energy from the sun to warm the water.

Urban planners must be creative. They must also be able to solve problems with many different people. Culhane describes his job as being like a hunter-gatherer. He says, "We hunt and gather new knowledge and solutions instead of food."

Thomas Culhane installing solar panels on a house.

Thomas Culhane uses computers to help him map out how resources will be used.

BECOME AN EXPERT

Geothermal Energy:
Using Earth's Heat Energy

Most of the energy on Earth originally comes from the sun. Geothermal energy is heat energy from within Earth. It comes from inside our planet, not from the sun. The temperatures deep inside Earth melt rocks and heat water. The temperatures are cooler closer to the surface. Sometimes the melted rocks and hot water make it to Earth's surface. The lava that pours from erupting volcanoes is melted rock. The hot water that reaches the surface forms hot springs and geysers.

Old Faithful, the most famous geyser in Yellowstone National Park, erupts when steam that was heated underground explodes to the surface.

Geothermal energy comes from inside Earth.

Mantle — •
Crust — •
Outer core — •
Inner core — •

TECHTREK
myNGconnect.com

Student
eEdition

Digital
Library

Geothermal energy is one of many energy resources people use. Most of the energy people use comes from fossil fuels, such as gasoline and coal. They contain **chemical energy**. Chemical energy is **potential energy** that is stored in substances. People burn them to turn the chemical energy into other types of energy. Fossil fuels are nonrenewable because they form over millions of years inside Earth. Geothermal energy is a renewable resource. You cannot use up the heat energy that is produced inside Earth.

We use energy to light our way at night and to get from one place to another.

chemical energy
Chemical energy is energy that is stored in substances.

potential energy
Potential energy is stored energy.

Geothermal Resources

Most places do not use much geothermal energy. Why? One reason is that different places have different amounts of geothermal energy. In most places, large amounts of geothermal energy are very deep underground. It would be difficult and expensive to reach it.

There are some places where the hot, melted rock is close to Earth's surface. These places have hot springs and volcanoes. California, Nevada, Hawaii, Utah, and other western states have the most geothermal resources. Many countries around the world use geothermal energy, including Iceland, Japan, Mexico, and the Philippines.

U.S.A. GEOTHERMAL RESOURCES

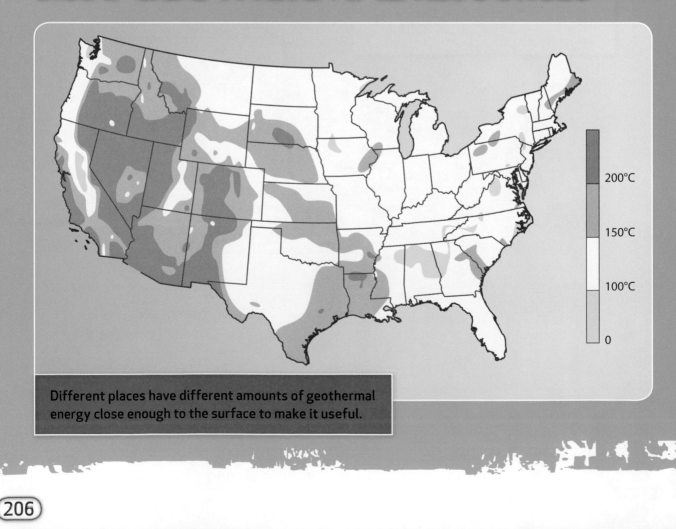

200°C

150°C

100°C

0

Different places have different amounts of geothermal energy close enough to the surface to make it useful.

People throughout history have used geothermal energy. They used hot springs to cook and bathe. Today, people use geothermal energy in many other ways. The heat energy in the hot water near Earth's surface is used to heat homes, swimming pools, and greenhouses. It is even used to provide warm water for fish farms. The energy in the water is also used to pasteurize milk and to dry onions, garlic, and tomatoes. The direct use of geothermal energy costs less than other ways of heating. It also produces very little pollution.

These shrimp grew up in a special pond heated by geothermal energy.

Workers harvest peppers in a greenhouse heated by geothermal energy.

Heating Buildings Iceland is an island nation that is close to the Arctic Circle. It has more than 20 active volcanoes. It also has plenty of geothermal energy. Most of the buildings in Iceland's capital, Reykjavik, are heated by geothermal energy. The buildings are heated by a district heating system.

This system uses hot water that is near Earth's surface. Wells are drilled to reach the water. Then, water that has a temperature of about 80°C (176°F) is sent through pipes. The pipes go to buildings throughout the city. Heat energy flows out of the hot water. The water becomes cooler. The air in the buildings becomes warmer.

Reykjavik is the world's northern-most capital city. However, warm ocean currents help keep the climate moderate. The average January temperature is −0.3°C (31.5°F). Many places in the United States are much colder than that.

Geothermal heat pumps are used to heat and cool buildings. In the summer, the pipes move heat from the building into the ground. In the winter, the pipes move heat from the ground into the building.

Geothermal heat pumps are very energy efficient and environmentally clean. They can also provide better heating and cooling than many common heating and air conditioning systems.

Geothermal heat pumps provide heat for buildings when it's cold outside. They can also cool buildings in hot weather.

Geothermal Power Plants

Geothermal energy can also produce electricity. Geothermal power plants use steam from hot water underground. In order to reach the water, a deep hole is drilled underground. The water is then pumped to the surface.

The steam from the water provides the **kinetic energy** needed to power a generator. A generator is a machine that can transform **mechanical energy** into electricity.

HOW DOES A GEOTHERMAL POWER PLANT WORK?

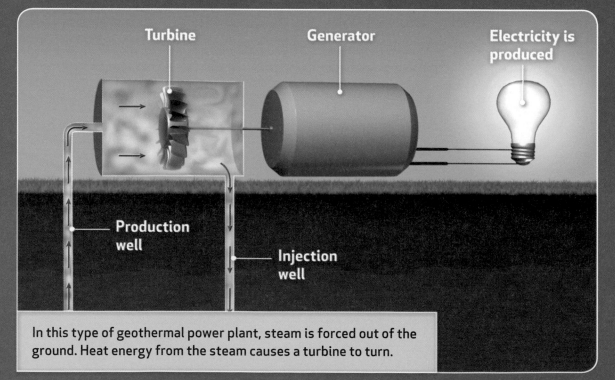

Turbine

Generator

Electricity is produced

Production well

Injection well

In this type of geothermal power plant, steam is forced out of the ground. Heat energy from the steam causes a turbine to turn.

kinetic energy
Kinetic energy is the energy of motion.

mechanical energy
Mechanical energy is an object's potential energy plus its kinetic energy.

The Geysers power plant in California has been running since 1960.

Earth has a supply of energy that can never be used up. Deep in the crust, water heated by hot rocks is a source of geothermal energy. Different places have different amounts of geothermal energy. People use geothermal energy to heat buildings and to make electricity. It is a renewable and clean source of energy.

TECHTREK
myNGconnect.com

Digital Library

The Blue Lagoon in Iceland offers geothermal seawater for swimmers. The water, which averages 40°C (104°F), is pumped up by the geothermal power plant. That would feel like a very warm bath, no matter what season of the year.

CHAPTER
5

SHARE AND COMPARE

Turn and Talk What are the benefits of using geothermal energy? Form a complete answer to this question together with a partner.

Read Select two pages in this section. Practice reading the pages. Then read them aloud to a partner. Talk about why the pages are interesting.

my SCIENCE *notebook*

Write Write a conclusion that summarizes what you have learned about energy. In your conclusion, restate what you think is the Big Idea of this section. Share what you wrote with a classmate. Compare your conclusions. Did you recall how energy can come in different forms?

my SCIENCE *notebook*

Draw Imagine a greenhouse that was heated with geothermal energy. Draw a picture that shows how pipes filled with geothermally heated water could heat a greenhouse. Add labels to show the direction in which heat would flow. Share your drawing with a classmate. Explain how the geothermal heat in your greenhouse is both the same as and different from the geothermal heat in your classmate's greenhouse.

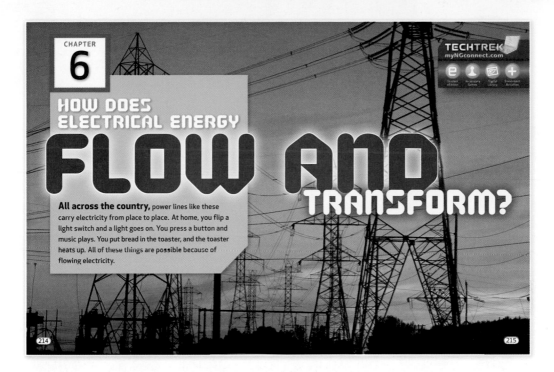

HOW DOES
ELECTRICAL ENERGY
FLOW AND
TRANSFORM?

All across the country, power lines like these
carry electricity from place to place. At home, you flip a
light switch and a light goes on. You press a button and
music plays. You put bread in the toaster, and the toaster
heats up. All of these things are possible because of
flowing electricity.

TECHTREK
myNGconnect.com

After reading Chapter 6, you will be able to:

- Explain that material that has been electrically charged attracts uncharged material
  without touching it and may either attract or repel other charged material. **ELECTRICITY**

- Classify materials that do and do not conduct electricity.
  ELECTRICAL CONDUCTORS AND INSULATORS

- Trace how electrical current travels by creating a simple electric circuit that will light
  a bulb. **ELECTRICAL CIRCUITS**

- Identify that electrical circuits provide a means of transferring electrical energy.
  ELECTRICAL CIRCUITS

- Understand that batteries store chemical energy that can change into electrical
  energy. **ELECTRICAL CIRCUITS**

- Explain how electrical energy can be produced from a variety of energy sources and
  can be transformed into almost any other form of energy, including light, heat, sound,
  motion, and magnetism. **ELECTRICAL ENERGY TRANSFORMS**

- Science in a Snap! Trace how electrical current travels by creating a simple electric circuit
  that will light a bulb. **ELECTRICAL CIRCUITS**

CHAPTER

6

HOW DOES ELECTRICAL ENERGY
FLOW

All across the country, power lines like these carry electricity from place to place. At home, you flip a light switch and a light goes on. You press a button and music plays. You put bread in the toaster, and the toaster heats up. All of these things are possible because of flowing electricity.

AND TRANSFORM?

TECHTREK
myNGconnect.com

Student
eEdition

Vocabulary
Games

Digital
Library

Enrichment
Activities

SCIENCE VOCABULARY

electricity (ē-lek-TRIS-it-ē)

Electricity is a form of energy that involves the movement of electric charges. (p. 218)

> These lanterns need electricity to give off light.

current electricity
(KUR-ent ē-lek-TRIS-it-ē)

Current electricity is a form of electricity in which electric charges move from one place to another. (p. 218)

> Current electricity carries power from these power lines to the electric appliances in homes.

static electricity
(STA-tik ē-lek-TRIS-it-ē)

Static electricity is a form of electricity in which electric charges collect on a surface. (p. 220)

> Static electricity attracts the girl's hair to the balloon.

my
Science
Vocabulary

circuit
(SIR-cut)

conductor
(kon-DUK-ter)

current electricity
(KUR-ent ē-lek-TRIS-it-ē)

electricity
(ē-lek-TRIS-it-ē)

insulator
(IN-sū-lā-ter)

static electricity
(STA-tik ē-lek-TRIS-it-ē)

TECHTREK
myNGconnect.com

Vocabulary
Games

conductor (kon-DUK-ter)

A **conductor** is a material through which electricity can flow easily. (p. 222)

Copper is one metal that is a conductor. Electricity flows easily through it.

insulator (IN-sū-lā-ter)

An **insulator** is a material that slows or stops the flow of electricity. (p. 222)

Plastic covers wires that conduct electricity. Plastic is a good insulator.

circuit (SIR-cut)

A **circuit** is a looped path of conductors through which electric current flows. (p. 224)

The wires, battery, and light bulb complete a circuit, which is why the bulb is lit.

217

Electricity

You probably know that to get lights such as the ones on this page to light up, you need to plug the cord into an outlet in the wall to get electricity to the lamps. You may know what it's like to sit in the dark when the electricity goes out. But what exactly is electricity? What causes it?

Electricity happens because of electrical charges. All matter is made up of particles that have electrical charges. Then why don't you feel electricity every time you touch something? Particles have positive charges and negative charges. In most matter, the number of positive charges is equal to the number of negative charges. That makes the matter neutral. That's why you don't feel any kind of charge when you touch it. Some matter is positive—the matter has more positive charges than negative ones. And some matter is negative—the matter has more negative charges than positive ones. So, what exactly is electricity? Electricity is the movement of negative charges.

Current Electricity
Most of the electrical machines that you are familiar with use current electricity. Current electricity flows along wires. The charges move through the wire so that you can light a lamp, toast bread, or listen to music!

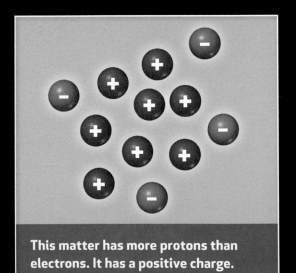

This matter has more protons than electrons. It has a positive charge.

This matter has more electrons than protons. It has a negative charge.

Electricity runs through wires
in these lanterns.

219

Static Electricity When you walk across a carpeted floor, your feet rub on the carpet. Then, when you touch something, such as a person, a pet, or a metal doorknob, you get a shock. What caused the shock?

That shock was caused by **static electricity**. Static electricity is a buildup of electrical charges on an object.

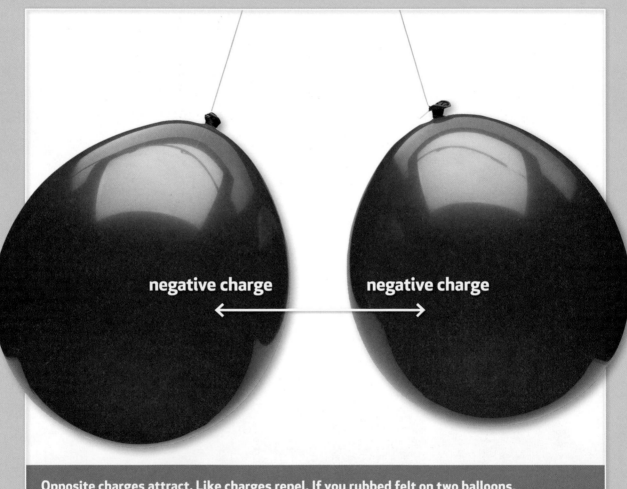

negative charge   negative charge

Opposite charges attract. Like charges repel. If you rubbed felt on two balloons, they would both pick up negative charges from the felt. If you brought the balloons close together, they would repel each other, because they both would have a negative charge.

Take a look at the photo below. What can you tell about the balloon and the girl's hair? You know that objects with opposite charges attract each other. If you rub a balloon on your hair, the balloon picks up negative charges from your hair. As those charges build up, static electricity builds, too. If you move that balloon near your hair, the negative charges on the balloon attract the positive charges on your hair. That's what makes hair "stick" to a balloon!

Static electricity can make things, like hair, move.

Before You Move On

1. What happens when two objects with like charges are near each other?
2. How is current electricity different from static electricity?
3. **Draw Conclusions** You place two balloons near each other, and they move away from each other. What might you conclude based on what you observe?

Electrical Conductors and Insulators

Electricity moves from one place to another. Think about a lamp again. Remember how the electricity moves from the wall socket through the wires in the cord to the light bulb? The wires in the cord are conductors. A conductor is a material that allows electricity to move easily. Metals such as copper, gold, silver, and iron are conductors. The wires in the lamp cord, for example, are copper.

Have you ever had to get out of an outdoor swimming pool because of lightning? That's because there are materials in water that make water a good conductor. If lightning struck the water, everyone in the pool would be shocked.

Because electricity can be dangerous, it is important to protect people from it. An insulator can do this. An insulator is a material that slows or stops the flow of electricity. Plastic, rubber, glass, wood, and ceramics are insulators.

CONDUCTORS

Copper wires carry electricity in power lines and electrical plugs.

Gold carries electricity in computer motherboards.

INSULATORS

Plastic covers electrical wires to keep electricity from shocking you.

Glass insulators on power poles prevent electricity from reaching a person working on the poles.

Wherever conductors are used, you'll find insulators as well. The gold circuits in a computer are mounted in plastic. Lights and pumps in a pool have plastic, rubber, and glass to separate the electric current from the water.

A telephone is completely covered in plastic and rubber to separate the person from the charge that makes the telephone work. Power lines have ceramic and glass insulators everywhere they connect to a pole.

Insulators cover the lights and the pumps in this pool. The lights and pumps are powered by electricity.

Before You Move On

1. What is a conductor?
2. What type of material would make a good insulator?
3. **Evaluate** Why are insulators important when working with electricity?

Electrical Circuits

When you run a complete lap around an oval track, you complete a **circuit**. A circuit is a line or a route that starts and finishes in the same place. Just like a runner on a track, electricity travels in circuits, too. Let's look at a circuit to see how this works.

A battery stores energy inside it. In a circuit, the battery's energy changes to electrical energy. One end of a battery, the one with the bump, is marked positive. The other end, which is flat, is marked negative. If you attach a wire from one end to the other, electricity flows from the negative end to the positive end of the battery. The wires create a circuit—a path through which electricity can flow.

For electricity to flow, a circuit must be closed. Think about that oval track again. If a big hole opened up in the track, you would not be able to complete your circuit. If an electrical circuit is open, the electricity cannot complete its circuit—it cannot flow. So, what happens when you turn a light on? You actually close a circuit, which allows electricity to flow from the source to the light you turned on. When you turn the light off, you open, or break, the circuit so that electricity cannot flow.

Enrichment
Activities

switch

light bulb

In this circuit, the electricity flows from the negative end of the battery, through the wire, into the light bulb, through the wire and paperclip, then back to the positive end of the battery. How can you open and close the circuit? Just move the paperclip like a switch!

Wrap the foil around the base of the bulb and clip the clothespin over the foil.

Tape the end of another piece of foil to the battery. Touch the foil strip to the bottom of the bulb to light it. Only touch long enough to light the bulb.

How did electricity travel through your circuit?

wire

battery

Before You Move On

1. In your own words, tell what a circuit is.
2. What happens to the electricity flowing through the wires in your wall when you flip the light switch?
3. **Infer** How would the circuit inside a flashlight work?

Electrical Energy Transforms

Look at the photograph of Barcelona's Magic Fountain. Lights glow. Water sprays. Music plays in time to the pulses of water. What makes all this happen? Electricity!

Electrical energy can transform, or change, into other kinds of energy. That's what makes electrical energy so useful. If you see light in the room, hear noise from a radio, or feel heat from a heater, you know how useful electrical energy is.

At Barcelona's Magic Fountain, electrical energy is changed into light, sound, and motion.

Take a look at the car below. Why do you think the car is plugged in?

The electric car is run by a battery. The battery provides the electricity. The electricity is converted into motion, heat, light, and sound. The driver plugs the car into a special outlet to recharge the battery.

The battery gives the car enough energy for a driver to warm up the seat, light the road ahead, and play favorite tunes on the radio!

TECHTREK
myNGconnect.com

An electric battery in the car creates light, motion, sound, and heat.

Digital Library

Heat At breakfast, you drop a piece of bread into the toaster. You push a button. In a minute or so, your bread is brown and toasted. But what really happened?

Inside the toaster, a special material slowed down the flow of electricity.

That caused some wires inside the toaster to heat up. The heat from the wires caused your bread to get brown. You might also have seen the wires glowing. The toaster transformed electrical energy to heat energy and light energy.

TECHTREK
myNGconnect.com

Digital Library

When electrical energy moves into the toaster, the electricity changes to heat and a small amount of light.

The chart below shows appliances in the home that change electricity into heat energy. Be careful, though! If these appliances get too hot or you feel heat from a machine that should not get warm, let an adult know. There may be a problem that could cause a burn or even a fire.

COMMON HOUSEHOLD APPLIANCES THAT MAKE HEAT

BLOW DRYER
Electrical energy transforms to heat energy and to energy of motion to create heat and blowing air.

OVEN
Electrical energy transforms to heat energy to cook food.

GRIDDLE
A griddle is another appliance that uses heat energy to cook food.

IRON
The heat of the iron smooths wrinkles in your clothing.

Light   You can probably name many kinds of objects that light up: ordinary light bulbs, fluorescent bulbs, neon lights, and even lasers. Each of these objects is made differently, but they all give off light.

Wires are good conductors. Electrical energy flows along wires to light bulbs. Inside a light bulb, there is a thin coiled wire called a filament. When electrical energy flows through the filament the flow of energy slows. This causes the filament to glow. You can see this in a light bulb because this energy change causes the light bulb to light up.

TECHTREK
myNGconnect.com

Digital Library

Lights shine all over Hong Kong. Some glow. Some glare. Some look white; some look blue or green. What do they all have in common? Moving electrical charges!

When the player strums the strings on an electric guitar, electrical signals are produced that affect the sound you hear.

Sound waves are coming out of this large speaker cone, causing the whole speaker to vibrate.

Energy of Motion

Many electrical machines and appliances have moving parts. This means that electrical energy transforms into the energy of motion. Electricity works in a special way to create the energy of motion.

Electrical energy transforms into motion and causes the head of this toothbrush to spin.

This electric personal transporter runs on electricity. The current from a battery runs a motor, which causes the wheels to spin.

Just about anything that changes electricity into motion contains a special part called a motor. As electricity flows through a motor, charged particles inside the motor start to move. They also change direction as they move. The constant change in direction causes a magnet inside the motor to spin. The spinning magnet makes other parts spin, too. The result of this spinning could be a toy that zips across the floor or a fan with spinning blades.

Look around your home. How many machines with electric motors can you find? A fan above a stove, a garbage disposal, and a turntable in a microwave all use electric motors. Ceiling fans, DVD players, and electric clocks do, too.

Some electric machines need more than one motor. Think about a desktop computer. One motor runs the fan to cool the equipment. Another causes the CD drive to open and close. Electric cars have many motors to control the wheels, the windshield wipers, the fan, sunroof, windows, and more.

An electric motor causes the blades of this fan to move.

Electrical Energy on the Job

You already know that electricity can be used to make objects move. Electricity can also be used to create magnets! A coil of certain kinds of metals with electricity flowing through it can create a magnet.

A magnet created by electric current is called an electromagnet. Electromagnets are temporary.

When electricity runs through the electromagnet, the electromagnet attracts some metals. As long as the circuit is closed, the electromagnet continues to attract the metal. When the circuit is opened and the current stops, the electromagnet drops the metal it is carrying.

Making a basic electromagnet is simple. You need a piece of iron, such as a nail, a battery, and some wire.

1. Wrap the wire around one piece of iron from one end to the other. Leave several inches of wire at each end.

4. Move a second piece of iron near the coiled wire. The coil is now magnetized and attracts the iron.

2. Attach one end of the piece of wire to the negative end of the battery.

3. Attach the other end of the wire to the positive end of the battery.

N   S

Battery — +

Electromagnets are extremely useful. Engineers can build very powerful electromagnets by using electrical current. These super-powerful electromagnets can lift large amounts of metal, which makes the process of moving loads faster. And, because electromagnets are temporary, machines can drop their heavy loads easily. The operator simply has to shut off the electrical current.

If the supply of electricity to the electromagnet were stopped, all these pieces of metal would drop from the magnet.

Before You Move On

1. Electricity transforms into other kinds of energy. What are they?
2. How does electricity create light?
3. **Infer** How can electricity cause a fan to move?

WHEN THE ELECTRICITY GOES OUT

On August 15, 2003, a surge of electricity moved through the power lines all over the northern East Coast of the United States. As it did, circuit breakers tripped and cut off electricity to homes and businesses. As the surge moved through the power lines, some power company employees shut electricity down in their towns and cities to keep the surge from reaching them. People call these surges "blackouts" because the lights go out.

The image on the left shows the lights on the U.S. East Coast that are visible from space. The image on the right shows roughly the same area after the blackout occurred.

A power surge is a sudden increase in the amount of electricity traveling in a current. A small power surge can cause damage to appliances and other electrical machines that are plugged in. A large power surge can damage power plants, power lines, and large electrical machines. In your home, a simple device called a surge protector can protect your appliances from being destroyed by a sudden increase in electricity. But a surge across an electrical system that supplies power to a city or an even larger area is a much bigger problem than a surge in your home. The power surge on August 15th affected more than single homes. When this surge was over, millions of people in the United States and Canada were left in the dark.

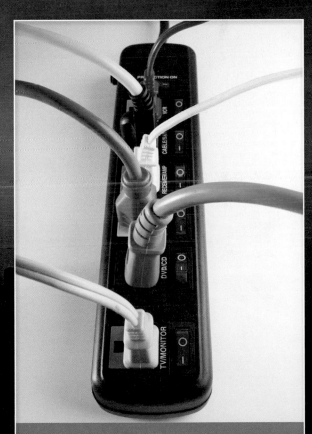

People plug their appliances into surge protectors like this one to protect against sudden increases in electricity.

New York City is the largest city in the United States. Power all across the city was shut off in order to protect all of the electrical equipment and power lines. But by turning off the power, everyday life in New York City came to a screeching halt. Thousands of people were stranded underground in dark subway cars. People working in high-rise buildings were trapped without an elevator, and the only way out of the buildings was to go down thousands of stairs in the dark. The people who were on the street had to battle with uncontrolled traffic.

All of the traffic lights were off. People all across the city were stranded, stuck in traffic jams, or forced to walk long distances home. The effects of the surge rippled across the United States and Canada, affecting travelers who wanted to fly from New York's airports.

Some power surges come from equipment that does not work properly, or people who use the equipment incorrectly. But power surges can also be caused by nature, such as a lightning strike. Power companies monitor the electricity flowing every minute of every day to try to stop surges before they cause damage.

Times Square, the famous and busy intersection in New York City, is normally flooded with lights.

With no electricity, the only lights in Times Square are from headlights and flashlights.

On August 15, no one was exactly sure what caused the blackout. Officials investigated whether terrorists caused the blackout, but found no evidence that any person had tampered with the electrical grid. Some blamed lightning. The President called for the electrical system to be updated. The blackout could have been caused simply by too many people using electricity to air condition their homes and offices on a hot summer day. Whatever its cause, the blackout showed that people rely on electricity in their everyday lives.

Workers restored electricity to the city, one section at a time.

Conclusion

Electrical charges move to create static and current electricity. Electricity flows easily through conductors, but not through insulators. A closed circuit allows current to flow, while an open circuit stops the flow of electricity. Electrical energy can transform into other forms of energy, such as heat, light, sound, and motion.

Big Idea Electrical energy flows and transforms into other types of energy.

ELECTRICAL ENERGY CAN BE TRANSFORMED INTO . . .

Heat   Light   Sound   Motion

Vocabulary Review

Match the following terms with the correct definition.

A. **insulator**

B. **current electricity**

C. **conductor**

D. **electricity**

E. **circuit**

F. **static electricity**

1. A form of energy that involves the movement of electric charges
2. A form of electricity in which electric charges collect on a surface
3. A form of electricity in which electric charges move from one place to another
4. A material through which electricity can flow easily
5. A material that slows or stops the flow of electricity
6. A looped path of conductors through which electric current flows

Big Idea Review

1. **Recall** What is electricity?

2. **Classify** Name three materials that are good conductors. Name three materials that are good insulators.

3. **Compare and Contrast** What is the difference between static electricity and current electricity?

4. **Explain** Choose one form of energy: heat, light, sound, or motion. In your own words, explain how electricity changes into that form of energy in a machine you have at home.

5. **Draw Conclusions** If a source of electricity is present, but a light bulb on a circuit is not lit, what can you conclude about the circuit? What would have to happen to make the light bulb go on?

6. **Describe** How would your life change if there were no electricity?

Write About Electricity Transforming

Explain What is happening in this photo? How is the electrical energy in the speaker transforming?

CHAPTER 6 PHYSICAL SCIENCE EXPERT: VIDEO GAME PROGRAMMER

The images, the movement, the sounds—they all have to come from somewhere. Video game programmer Tara Teich puts them together so that video games "come to life" for game players.

Tara Teich

NG Science: What is your job title, and what do you do?

TT: I'm a video game programmer. I spend my days working at a computer writing the code that brings games to life. My job is to help bring the design, art, and sound together to create gameplay. I help make characters walk when you move the controller and jump when you hit a button. I help make them talk when they see something interesting!

NG Science: What is a typical day at work?

TT: I spend a lot of the day at my desk writing code. But I spend an equal amount of time talking to other engineers. We discuss what they are working on and what is the best way to solve a particular problem. I discuss how the game works with the designers, and I figure out how to make everything more fun.

Student
eEdition

Digital
Library

NG Science: What did you have to do to become a video game programmer?

TT: I have a college degree in computer science. This is a degree that would let you be a programmer for any company, not just for a video game company.

NG Science: When did you know that you wanted to be a programmer?

TT: I started programming when I was around 11, and I knew I loved it even then. Being able to bring anything I could imagine to life on the screen was very exciting.

NG Science: What do you like best about your job?

TT: The coolest thing about programming is that there is not just one solution to a problem. I like that my work is limited only by my imagination. I can write code to do anything that I can think of, if I can puzzle out how to do it.

Digital
Library

Teich uses multiple computers as she programs new games.

243

BECOME AN EXPERT

Video Games: When Electricity Becomes Really Fun

Video games are a huge part of today's culture. Although some parents may doubt their value, kids all around the world use them as a prime source of fun. The thousands of games available today started out as research projects. Professors and students wanted to show what they could get **electricity** to do.

In 1952, a student at Cambridge University created a tic-tac-toe game to run on a university computer. In 1962, students at MIT were able to use the new, smaller computers to create a game called *Spacewar!* It became a hit at colleges all over the country. Never heard of it? That's because the computers needed to play it would take up entire rooms. Only colleges and research facilities had the equipment to play the games.

It took a machine large enough to fill a room to play the simple game of *Spacewar!*

electricity

Electricity is a form of energy that involves the movement of electric charges.

244

TECHTREK
myNGconnect.com

Student
eEdition

Digital
Library

So what do these old computer games have in common with the game systems people use today? All of these games are run by computers.

Computers use **current electricity** to perform instructions. Instructions might be calculations, directions for the cursor to move, or a command to beep. In the video game systems you play today, the directions include making images appear on the screen, sending music and other sounds through the speakers, and sending vibrations to your controller.

current electricity

Current electricity is a form of electricity in which electric charges move from one place to another.

Inside the video game system is a computer that works with the television and the controllers to create a fun and exciting game.

245

Inside every game system is a microchip. One microchip holds many **circuits** . The circuits are too small to see with the naked eye. The microchip controls what the video game does.

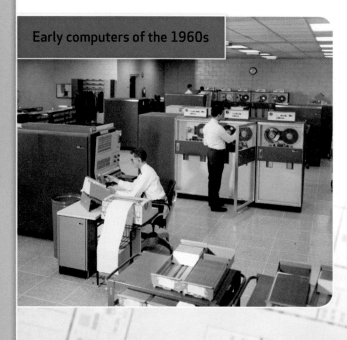

Early computers of the 1960s

If there's an explosion, you'll see it and hear it. And the microchip is also in charge of awarding or taking away points.

The microchip does not know these directions on its own. Each game has code, or instructions, for the game system. This is why you have to change game disks or cartridges to go from one game to the next. Some game systems store some game codes for you. Then, you simply tell the system to load a game. In other words, it goes out to find the correct code.

The microprocessors that game systems use today are much smaller, but far more powerful, than those in the early computers of the 1960s.

circuit
A **circuit** is a looped path of conductors through which electric current flows.

The microprocessor inside any gaming system or computer is attached to a larger set of circuits, called a motherboard. The motherboard is usually made of plastic with copper paths pressed into the surface. The plastic works as an **insulator**. It keeps the electricity that operates the game from reaching people or things it should not reach.

The copper paths on the motherboard create many circuits. These control all of the machinery— the television screen, the controller, and so on. Basically, the motherboard sends out all of the instructions the microprocessor gives. Copper is used on the motherboard because it is an excellent **conductor**.

TECHTREK
myNGconnect.com

The motherboard of any video game system or computer is a thin plastic board with circuits of copper pressed into the surface.

Digital Library

insulator

An **insulator** is a material that slows or stops the flow of electricity.

conductor

A **conductor** is a material through which electricity can flow easily.

The difference in the systems is the type of computer equipment inside. Each type of game system reads directions in a particular way. It is as though Game A speaks Spanish and Game B speaks Swahili. They just don't understand each other's directions. This explains why you cannot put one game into another game's play system and expect it to work.

This means that if the same video game is available for two game systems, then two entirely different sets of instructions had to be written. The programmers writing the code must understand how both game systems receive and send instructions. The time it takes to write the code can cost the company a great deal of money. This is why many games are available for only one system.

Video games are fun. Believe it or not, they are also useful! People who play games want great graphics. The more things a video game system can do, the better. Many improvements that people have made to computers happened with video games first. People who work with "regular" computers borrow technology from the gaming technology. The same graphics cards that make games fun can work for business, too.

A video game engineer "builds" images for a new video game. He also writes code so that the game runs according to design.

The games themselves can help people on the job. The U.S. military uses video games to teach spies how to think and react. Doctors use video games to help patients getting treatment for cancer.

NASA has even developed technology that helps kids with attention deficit disorder (ADD) use video games to improve their focus.

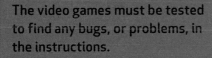

The video games must be tested to find any bugs, or problems, in the instructions.

It's interesting to know about the workings of video game systems. However, that does not mean you should take apart any of the pieces of your game systems.

Do not open up your game systems if you want them to work. As interesting as it is to look inside and become more of an expert, you would very likely cause a problem. You run the risk of passing **static electricity** on to the motherboard or other parts of the computer. This static electricity can damage the circuits and cause the game not to work.

Why is it so important to take care of game disks and cartridges? Dust and scratches can damage the code. The computer will no longer understand it. Then the game won't work.

Taking care of your games will help them last much longer.

static electricity

Static electricity is a form of electricity in which electric charges collect on a surface.

A game system controller such as this is just one part of the system that receives instructions from the motherboard.

SHARE AND COMPARE

Turn and Talk How does electrical energy change into other forms of energy in video games? Form a complete answer to this question together with a partner.

Read Select two pages in this section. Practice reading the pages. Then read them aloud to a partner. Talk about why the pages are interesting.

my SCIENCE notebook

Write Write a conclusion that tells the important ideas you learned about video games. State what you think is the Big Idea of this section. Share what you wrote with a classmate. Compare your conclusions. Did your classmate make the connection between video games and electrical energy?

my SCIENCE notebook

Draw Someone asks you, "How does a video game use electrical energy?" Draw a diagram to answer the question. Combine your drawing with those of your classmates to make a video game "How Does It Work?" guide.

Glossary

A

atom (A-tum)
An atom is the smallest piece of matter that can still be identified as that matter. (p. 18)

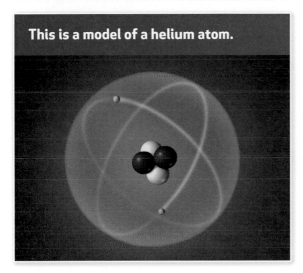

This is a model of a helium atom.

C

chemical change (KEM-i-kul CHĀNJ)
A chemical change is a change in matter that forms a new substance with different properties. (p. 68)

chemical energy (KEM-i-kul EN-ur-jē)
Chemical energy is energy that is stored in substances. (p. 194)

circuit (SIR-cut)
A circuit is a looped path of conductors through which electric current flows. (p. 224)

condensation (kon-din-SĀ-shun)
Condensation is the physical change of matter from a gaseous state to a liquid state. (p. 66)

conductor (kon-DUK-ter)
A conductor is a material through which electricity can flow easily. (p. 222)

current electricity (KUR-ent ē-lek-TRIS-it-ē)
Current electricity is a form of electricity in which electric charges move from one place to another. (p. 218)

E

electricity (ē-lek-TRIS-it-ē)
Electricity is a form of energy that involves the movement of electric charges. (p. 218)

evaporation (ē-va-pōr-Ā-shun)
Evaporation is the physical change of matter from a liquid state to a gaseous state. (p. 64)

The water droplets on the side of the glass are condensation.

F

force (FORS)
A force is a push or a pull. (p. 98)

G

gravity (GRA-vi–tē)
Earth's gravity is a force that pulls things to the center of Earth. (p. 106)

I

inclined plane (IN-clīned PLĀN)
An inclined plane is a flat surface with one end higher than the other. (p. 142)

insulator (IN-sū-lā-ter)
An insulator is a material that slows or stops the flow of electricity. (p. 222)

The plastic that covers these wires is a good insulator.

K

kinetic energy (ki-NET-ik EN-er-jē)
Kinetic energy is the energy of motion. (p. 180)

L

lever (LE-vur)
A lever is a bar that turns against an unmoving point. (p. 141)

M

mass (MAS)
Mass is the amount of matter in an object. (p. 14)

mechanical energy (mi-KAN-i-kul E-nur-jē)
The mechanical energy of an object is its potential energy plus its kinetic energy. (p. 180)

mixture (MIKS-chur)
A mixture is two or more kinds of matter put together. (p. 26)

motion (MŌ-shun)
Motion is a change in position. (p. 98)

The force of the ball hitting the cans puts the cans in motion.

P

physical change (FI-si-kul CHĀNJ)
A physical change is when matter changes to look different but does not become a new kind of matter. (p. 59)

potential energy (pō-TEN-shul EN-er-jē)
Potential energy is stored energy. (p. 180)

pulley (PŪL-lē)
A pulley is a grooved wheel with a cable or rope running through the groove. (p. 150)

S

screw (SCRŪ)
A screw is a bar that has an inclined plane wrapped around it. (p. 144)

solution (so-LŪ-shun)
A solution is a mixture of two or more kinds of matter evenly spread out. (p. 30)

static electricity (STA-tik ē-lek-TRIS-it-ē)
Static electricity is a form of electricity in which electric charges collect on a surface. (p. 220)

V

volume (VOL-yum)
Volume is the amount of space something takes up. (p. 16)

W

wedge (WEJ)
A wedge is a double inclined plane that can split objects apart. (p. 146)

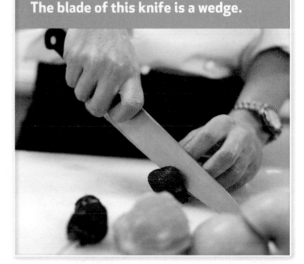

The blade of this knife is a wedge.

wheel and axle (WĒL AND ACK-sel)
A wheel and axle is a large outer wheel attached to a smaller wheel called an axle. (p. 148)

This swimming pool contains a solution of chlorine and water.

Index

Credits